ENGINEERING
AND
ENGINEERS

DE DIVERSIS ARTIBUS

COLLECTION DE TRAVAUX
DE L'ACADÉMIE INTERNATIONALE
D'HISTOIRE DES SCIENCES

COLLECTION OF STUDIES
FROM THE INTERNATIONAL ACADEMY
OF THE HISTORY OF SCIENCE

DIRECTION
EDITORS

EMMANUEL
POULLE

ROBERT
HALLEUX

TOME 60 (N.S. 23)

BREPOLS

PROCEEDINGS OF THE XXth INTERNATIONAL CONGRESS
OF HISTORY OF SCIENCE (Liège, 20-26 July 1997)

VOLUME XVII

ENGINEERING AND ENGINEERS

Edited by

Michael Ciaran DUFFY

BREPOLS

The XXth International Congress of History of Science was organized by the Belgian National Committee for Logic, History and Philosophy of Science with the support of :

ICSU
Ministère de la Politique scientifique
Académie Royale de Belgique
Koninklijke Academie van België
FNRS
FWO
Communauté française de Belgique
Région Wallonne
Service des Affaires culturelles de la Ville de Liège
Service de l'Enseignement de la Ville de Liège
Université de Liège
Comité Sluse asbl
Fédération du Tourisme de la Province de Liège
Collège Saint-Louis
Institut d'Enseignement supérieur "Les Rivageois"

Academic Press
Agora-Béranger
APRIL
Banque Nationale de Belgique
Carlson Wagonlit Travel - Incentive Travel House

Chambre de Commerce et d'Industrie de la Ville de Liège
Club liégeois des Exportateurs
Cockerill Sambre Group
Crédit Communal
Derouaux Ordina sprl
Disteel Cold s.a.
Etilux s.a.
Fabrimétal Liège - Luxembourg
Generale Bank n.v. - Générale de Banque s.a.
Interbrew
L'Espérance Commerciale
Maison de la Métallurgie et de l'Industrie de Liège
Office des Produits wallons
Peeters
Peket dè Houyeu
Petrofina
Rescolié
Sabena
SNCB
Société chimique Prayon Rupel
SPE Zone Sud
TEC Liège - Verviers
Vulcain Industries

D/2002/0095/51
ISBN 2-503-51409-x
Printed in the E.U. on acid-free paper

TABLE OF CONTENTS

TABLE OF CONTENTS

History, Philosophy and the Changing Nature of Engineering

Michael Ciaran DUFFY

INTRODUCTION

The changing nature of engineering demands a new approach to the history and philosophy of engineering. Recent innovations present analysts with an inter-related set of historical, philosophical and engineering problems which must be solved if continued engineering progress is to be facilitated. Present-day strategic engineering design compels engineers to create an internalist history, and an internalist philosophy of engineering for themselves, granted the neglect of internalism by most historians and philosophers of technology, whose goals and priorities are different, and whose engineering skills are inadequate for this task.

History and philosophy of engineering are distinct from history and philosophy of technology as the latter are generally understood in the 1990s, though they can be related. No adverse criticism of historians and philosophers of technology is implied. The author has a different aim from them, which is to confront problems in engineering requiring the methods of historical, philosophical and engineering analysis for solution. His goal is to serve engineering by analysing the changing nature of those strategic components and systems which define what is meant by exemplary technology in the late 21st century.

STRATEGIC INNOVATION

Strategic engineering innovations set new standards for determining what is meant by modern engineering at its best. They introduce the engineering components and systems on which new industries are founded.

During each era, relatively few industries set the global standards, and lead industrial and economic growth. Each of these industries depends on a particular engineering system, or combination of systems, which represent the best practice for that time. These new engineering systems deepen, broaden and transform contemporary understanding of what engineering is, with consequences for industry, education and culture. Of particular importance are those innovations which result in a period of global economic growth. Identifying the strategic engineering innovations in different periods, and different cultures, is a major task for historians.

In each period there are artefacts, systems, theories, and methods which are associated with technologies which originated in different periods. Amongst these will be those devices and methods which call into existence new industries, with new standards of training, education, funding, organisation, management, production, construction, and research. They introduce new ways of thinking about engineering and its physical and cultural environments, and they cannot be understood without directing attention to the conceptual apparatus, symbolism, and language by which they are defined, analysed and described.

Today, in such fields as artificial intelligence, virtual reality engineering, and engineering studies of perception, the abstract conceptual apparatus is as much an engineering artefact as the tangible, material equipment. Engineering innovation is creating new concepts and models for interpreting human experience and physical phenomena and engineering has become a creative science second to none.

Most carefully classified analytical studies of engineering innovation concentrate on the post-1600 period, in part because they have been carried out by innovation analysts and economists seeking a deeper understanding of the world built by mechanised industrial-capitalism. Much needs to be done on earlier periods to compile detailed histories of the engineering innovations during the Classical and Medieval epochs in Europe, and in the great civilisations of North Africa, the Near East, and Asia. By its very nature, an innovation has to be perceived as such.

It is a recognisable " event " or development, which becomes easier to define, and analyse after 1600, when relatively rapid changes in equipment form, within the life of an engineer, enable innovation to be identified as a distinct activity. It is questionable whether one can refer to innovation in the ages when mankind's ancestors were evolving, and learning to use simple implements without the philosophy, language and ideas to understand them.

A much deeper insight will be gained into the evolution of engineering when more is known about developments in the pre-written language period, and when more has been done on the first correlations between technical developments and corresponding developments in language, symbolism and philosophy.

Through the lack of such studies, too much attention has been directed onto " engineering as objects " rather than onto " engineering as ideas ".

Exemplary Innovations

Identifying the exemplary innovations and the standards-setting technical systems which are found in the great civilisations enables the path of engineering progress to be traced through a thousand years of complex social and economic change. Engineering exemplars may be likened to paradigms, a concept which originated in grammar, and which was used by C.G. Lichtenberg in the 18th century, and which was developed in the 1960s by T.S. Kuhn in his study of conceptual revolutions in science. Studying the rise and fall of strategic technologies, and their influence on economic growth, shows that the nature of engineering artefacts and engineering method has changed considerably since 1600.

The 17th century can be taken as being the period when scientific engineering, as an intrinsically progressive, distinct activity came into being. Completely new engineering achievements, such as the steam engines and powered textile mills of the 18th century, surpassed those of former civilisations and implanted the idea of technological change. No longer did centuries-old designs, taken from drawings of Roman machinery, continue to guide engineers more than twelve hundred years after the Empire fell.

In the mid-18th century, engineers expected noticeable improvement in their lifetimes. In the 19th century, constant improvement even in conservative engineering was taken for granted. Today, in microelectronics, computing, robotics, nanotechnology, and other exemplary fields, major problems are caused by the need to keep pace with innovation, and to integrate advances made in several disciplines. Modern engineering can now transform itself rapidly.

There is insufficient space here to list the more important exemplary innovations in the history of engineering from circa 250,000 BC. Any such attempt raises the question as to what is meant by engineering in the periods before language, systematic thinking, and systematic symbolism. The stone axe of 250,000 BC is an artefact made by early man, but how was the process of manufacture recognised, remembered, co-ordinated with mental activity ? When did manufacture become a conscious, recognised activity ? An exemplary device from much later, circa 3500 BC, such as a four-wheeled steppe wagon, or the spoked-wheeled Chinese chariot of *c.* 1000 BC raise the question of how to assess the ability of their builders to imagine devices ; to criticise them in the imagination ; to idealise them ; to conceive of a better design after which they could strive.

Engineering history will remain defective until studies of the use of symbols, ideas and imaginary models in technics are carried out. This will demand going far beyond the activities thought proper for most " history of engineering " societies at the present time. The author believes that what we mean by modern engineering method, with its integration of imagination, idealisation, and insight gained by rational observation, analysis and experiment, becomes a major source of innovation and industrial change in the 17th century, though it is discernible much earlier. Identifying some of the exemplary innovations in each century witnesses to the changing nature of engineering.

The man-powered organ bellows ; the 15th century Duringer clock ; the water-driven fulling mill ; the 17th century Savery steam engine ; the 18th century Newcomen engine ; the 19th century steam locomotive ; the 19th century telephone, electric motor, or power station ; the 20th century wireless telegraph, broadcasting system, totalisator, computer-controlled signalling network, electronic code breaking system, virtual reality system, or microprocessor implanted into a cockroaches back : each represents best practice in a particular culture in a particular age. As we approach the present, the devices depend more and more on imagination, abstraction, and the use of evolving conceptual apparatus, and analytical tools, for their creation. Their span of usefulness shortens, and calls for historians able to deal with the events of the last five years. This challenge will hardly appeal to those historians who turn to history, including history of engineering, to escape from the present.

The role of innovations in the industrial-capitalist epoch can be briefly summarised as follows. The Newcomen atmospheric steam engine of the early 18th century marked the start of a mechanised industrial society which evolved through innovation, and accelerated the pace of its own expanding development, though admitting this does not deny the innovative achievements of earlier times. There have been distinct phases of growth in the global economy dominated by engineering change, and its consequences. Economists argue that each distinct phase is begun by engineering innovation. The Newcomen engine, the Crompton " mule ", the Darby coking oven, and water-powered machinery, transformed the industrial process and enabled it to expand. These strategic innovations created the strategic industries during the first period.

During periods of growth, a limited number of industries, based on engineering innovations of equal degrees of modernity, interact and stimulate each other. They foster exchanges of ideas, practice, products and staff. Growth industries serve as markets for other growth industries. Sometimes this group of strategic industries may be dominated by one — the " grand exemplar " — which best of all others exemplifies the vision, science, technology, organisation, quality of workforce, research initiative and other qualities which the others emulate. The influence of this group extends far beyond itself to the many small industries and businesses which depend on it ; to the worlds of education and training ; to commerce ; to public administration and politics ; and to popular culture.

During growth periods, available funds flow into these industries, but a time comes when they become less fruitful, perhaps because their technologies become obsolete. The industry itself might continue to be economically important, as was the British textile industry after 1890, without being a source of strategically significant innovations of a world-transforming class. Such industries are vulnerable to any better equipped, better organised modern rival. Available capital, seeking maximised returns should be moved into the innovations which are creating the strategic industries of the next phase. The geographical location of the strategic industries often shifts from epoch to epoch.

INNOVATION & INDUSTRIAL GROWTH

One common classification of the main epochs is as follows.

The first period was created by the engineering innovations in the textile, coal and iron industries, and was caused by using water and steam power to drive automatic and semi-automatic machinery which greatly increased output per worker. During this period, perhaps the most important strategic activity was building up expertise in application of steam power. This textile, coal and iron " revolution " took place in Great Britain which enjoyed an expertise in strategic skills not easily overtaken by rival countries at that time.

The second phase was dominated by the steam railway, with its attendant industries and services, which generated new ideas, methods and services which further transformed society. Because many of the strategic technologies of this " Railway Age " evolved from those found during the first period, Great Britain continued to be the leading industrial nation.

The third period began with the advent of heavy-duty industrial electrification, dominated by traction, the growth of the motor-car industry, and the introduction of industries and services dependent on systematic research : wireless telegraphy, chemicals manufacture and aviation. The engineering initiative, and leadership in strategic industries passed to the United States and Germany.

A fourth phase was introduced by the innovations of the 1939-1945 war and its immediate aftermath : electronics, advanced aviation, rocketry, nuclear power, computing and telecommunications.

A fifth era has been introduced by nanotechnology, advanced bioengineering, very large scale integrated electronics systems, information technology, advanced robotics, virtual reality systems and artificial intelligence systems.

Engineering studies of the brain, mind and consciousness promise further revolutionary and fundamental changes, not least in mankind's self understanding. In each of these epochs, the strategic innovations, and the industries founded on them, changed contemporary understanding of engineering. The Arkwright water frame (a water-driven machine for spinning threads), and the Crompton " mule " (a hybrid machine for cotton spinning), became major components in the integrated factory system, and were mechanisms which represented a new order of complexity and productivity. The steam-railway developed the concept of machine-ensemble in the systems context, and exemplified the large, influential, financially powerful public institution dominated by private capital. The electric power industry introduced the large, integrated, quan-

tifiable system which required advanced scientific techniques for its construction, for its daily operations, and for anticipating the future demands placed on
it. Electrification was widely hailed as marking a transformation in the nature
of history and civilisation. In the post-war era, the electronic computer has
transformed virtually every activity. Further developments based on electronics
and new materials have introduced nanoengineering, artificial intelligence,
machine intelligence, and virtual reality systems which have created new
industries, dependent on a new kind of engineer.

The engineering innovations demand innovations in education, training and
professional organisation. Attitudes to engineering, the profession, and to education and training date from the second and third epochs mentioned above,
and were defined through a response to systems which are no longer strategic
and exemplary.

Developments in the fourth epoch have led to considerable changes in the
engineering profession, and in education, but continuing radical transformations will be required to accommodate the innovations of the coming century
when the strategic engineering systems may result from integrating neuroscience, evolutionary genetics, computer science, microelectronics, nanoengineering, molecular physics and quantum mechanics. These innovations will
transform public understanding of engineering, and will create industries very
different from those which dominated global economics in the past.

ENGINEERING METHOD

The nature of engineering cannot be understood unless engineering method
is defined, and unless conceptual apparatus, and methods of analysing engineering systems, are considered equally with the disposition of components
within an engineering ensemble. Putting it simply, engineering method is a
particular case of scientific method, which thereby involves the creation of
concepts, models, and methods based on observation and checked by experiment, which relate precepts and correlate sets of ordered sense-data. These
concepts are employed to interpret idealised systems which can be analysed in
thought-experiments prior to being used to guide the design and construction
of some concrete equipment or system. Experience with the actual, realised
system leads to its performance being observed, measured and modelled. The
model of actual performance is compared with the model of expected performance, which is probably different from the model of ideal performance. A
model of improved performance is created, which in turn guides a second real-
isation of a better actual device, and so on, with successive approaches to the
ideal until an actual device, good enough for commercial and industrial success, is obtained. This creation of various sets of model — the experimental
based on experience and observation — and the ideal — based on imaginative
thought-experiment — are associated with an ongoing comparison between the

two sets of model. An imaginary system, abstracted from the actual, is thereby progressively perfected by approximating to an ideal, and is repeatedly used to improve design and production of better realised types : the actual machines. But the ideal is not static, and it evolves as knowledge increases.

All engineering creations have to fit into a larger " receiving system ". Each original design is conceived within a defining envelope, and there needs to be a model of the interface with the receiving system into which the actualised technology must fit. This receiving system changes all the time. It may be a radio broadcasting system ; a system of food production ; a telecommunications network ; a marine engine ; a microprocessor component, or a piece of domestic equipment, but the new component (which may be a system in itself) must interface with it, and the connecting relations have to be defined at outset with sufficient clarity to ensure that the component will function efficiently in its environment.

The overall process of design and integration involves fitting two systems together, the engineering component, and the receiving system, each of which changes in ways requiring endless modelling. Some components and systems can be described almost entirely in quantifiable, engineering terms, but difficulties arise when the design embraces vast systems where it is not possible to assign parameters to the defining envelope or the interface between component and receiving system. This is particularly the case where system parameters (such as fuel price, or labour costs) are subject to non-predictable changes. However, as engineering has developed, the analytical methods, and the conceptual apparatus for detailing and quantifying the above process, have become more extensive and in some cases more accurate, especially in research and development, and in narrowly technical environments. Despite ambitious programs in the USA, and to a lesser extent in Europe, attempts to model the general economic receiving system into which new technologies fit have not been successful.

ENGINEERING ANALYSIS & CONCEPTUAL APPARATUS

The changing nature of engineering is displayed in the increasing importance of the analytical tools ; the conceptual apparatus ; and the techniques for modelling actual, abstract and ideal engineering systems. The engineering systems which were exemplary in former ages could often be constructed without an integrated system of abstract models, supported by mathematical analysis, scientific experiment and quantification.

Examples include Medieval bell founding and organ building ; 17th century military engineering ; and much 18th century civil engineering — like lighthouse building. The construction of canals and wagon ways in the 18th century ; and the construction of railways before 1830 did not demand integrated systems of mathematical analysis, though they demanded — and got —

a great deal of Enlightenment science and reason applied to industrial problems. But after 1870, there is a marked increase in the perfecting of engineering method as defined above. Without very advanced mathematics, there could
have been no industrial electrical engineering ; no commercial wireless telegraphy or broadcasting ; no aviation. Without science, and rational analysis,
there could be no chemicals industry ; no mass production ; no modern materials. The trend is towards increasing reliance on analytical tools ; modern
mathematics ; abstract modelling and thought experiment, which suggest new
technologies, and indicate how links might be established between departments
of engineering, science and other disciplines once thought quite disconnected.
There is no longer an excuse for thinking of engineering in terms of matter
alone : it is now, in its exemplary departments, much more a matter of mind.

Engineering always disclosed relationships between abstract concepts, and
many devices were created through efforts to embody in mechanism properties
which were metaphysical, occult, or abstract. Shifting the goal from occult to
rational-abstract underlies engineering design as it evolved between the 13th
century and the 19th century.

In the 13th century, it was common to try to reproduce in mechanism an
ideal attribute observed in nature, such as the perpetual motion of the spheres.
The gradual separation of metaphysics, alchemy and the rational-physical led
to the scientific study of mechanisms which modelled physical actions
mechanically, though ridding science and engineering of occult notions was a
very slow business.

In the 17th century, a rational, mechanical philosophy triumphed to exercise
a general influence throughout mechanised, industrial culture. In a brief paper
such as this one or two examples must suffice. The evolution of musical instruments, and the design of clockwork, embodied metaphysics and philosophy,
though few histories of engineering bring this out.

Museums nearly always display machinery in terms of visible actions rather
than expressions of ideas. Consider the astronomical clocks, built by Hans
Duringer, such as the one being restored in St. Mary's Church, Gdansk,
Poland. Built between 1464 and 1470, this device is a philosophical machine
which replicates in mechanism the ideal motion of the celestial bodies. Much
of 18th and 19th centuries engineering development was profoundly influenced
by Enlightenment philosophy, as was recognised at the time. Lazare Carnot
(and F. Reuleaux) analysed mechanisms in terms of the fitness and harmony of
components and constructs which mirrored the harmony of the physical order.
The imaginary, idealised heat engine of Sadi Carnot, is an example of a disclosing model, which established a relationship between the imaginary and
ideal, and the actual and imperfect. Thought-experiments conducted with it
helped to show the ways by which designs could be improved. Thought-experiments conducted with an imaginary engineering disclosing model, abstracted
from the actual piston-and-cylinder atmospheric engines of Huygens and

Papin, enabled Brillouin to investigate the Maxwell " Sorting Demon ", and to establish a relationship between entropy and information exchange.

The role of engineering devices, many of them found in industry, in stimulating the imaginary disclosing model has been almost totally ignored, yet imaginary devices, suggested by industrial equipment, have led to great advances in conceptual apparatus. In order to improve electric motor design, and solve network problems, G. Kron created an idealised universal motor, and a system of analysis which relied heavily of imaginary entities, and resulted in an ambitious scheme for geometrising engineering. He did much to introduce tensors into engineering, and devised a general method of analysing and ensemble with mechanical or electrical elements, which he termed *Diakoptics*. Kronts work, and the vast conceptual apparatus demanded by electrical engineering is seldom featured to any extent in histories of electrical engineering. It is admittedly difficult to link the conceptual to the actual, for instance in a museum, but unless the effort to do so is made, the simple narrative history, or the museum display, are incomplete and possibly grossly misleading.

The management, organisational and production theories of F.W. Taylor, H.L. Gantt, F.W. Gilbreth, and H. Ford sought to incorporate in industrial activities the order and harmony which 19th and early 20th centuries science found in the physical universe. The work of Taylor and other pioneers of rational or scientific management mark the rise to importance in engineering of theories of organisation which became as important as theories of materials failure.

Exemplars in Modern Engineering

Current developments show how much engineering has changed since the 1960s when " best practice " was represented by a Saturn rocket, the Dounreay Fast Breeder Nuclear Reactor, the early satellites, or the first solid-state electronics systems. The work of Professor Cochrane and his team, at the Martlesham Heath laboratories of British Telecommunications, suggests that in the 21st century it may be possible to interface silicon chips with the human brain, perhaps by developing the equivalent of nerve endings on the chips. If this proves possible, the carbon-based memory systems of the biological brains, which have evolved on Earth, could be linked to the silicon-based systems of information technology. Perhaps the capacity of the human brain could be considerably increased sufficiently to mark a discontinuity in the evolutionary progress of mankind and herald the advent of a radically new bio-technical ensemble.

The developments in virtual reality engineering, artificial intelligence, and other modern disciplines require for their understanding a considerable degree of philosophy in addition to some familiarity with cognition studies, neuroscience, psychology and information technology. The response by engineering schools will need to be profound and extensive. The work of Prof. J.O. Gray into advanced robotics at Salford University uses technologies for simulating the sensed movements through perceived but imaginary environments. Imaginary surfaces can be touched using the Uttal Glove. Imaginary environments can be visualised and explored. Experience with these innovations raises questions concerning the ontological status of precepts, concepts, mental constructs, visual images, imaginary and actual entities. It prompts users to ask what is meant by reality, and how it is defined and recognised. These new systems suggest new ways by which human perception of the spatio-temporal environment might be interpreted.

The virtual reality systems, which simulate imaginary environments, stimulates questions from students about matters which out of date educators regard as the concern of philosophers only. Today, questions concerning consciousness, perception, ontological status of the perceiver and the perceived, and the meaning of " reality " are asked by engineers working in advanced areas of their discipline. Engineers routinely study intelligence, the nature of mind and the nature of life (artificial and biological) to improve further the fields of control, robotics, neuroscience and artificial intelligence. Artificial Life is one of the latest subjects to be listed in the publications of Oxford University Press, where the proceedings of the Tufts Symposium examines the relationships between Darwinism and Artificial Intelligence, drawing on Biology, Cognitive Science, Mathematical Genetics, Philosophy of Science and of course AI.

In other areas engineers are examining the intelligence and control systems of insects and other life forms, to assist the developments of control systems in robots. The analysis of Artificial Intelligence and Artificial Life demands a comparison of different kinds of life form, and different levels of intelligence as they have evolved. Links between once-separated disciplines are creating new disciplines and changing the nature of engineering in the late 20th century. In the last year, a Japanese mechanical engineer removed the wings from cockroaches and implanted microprocessors in their backs to control their movements. Telecommunications engineers have interfaced a rat's brain with a microprocessor and monitored its neuronal activity — using electron microscopy to take pictures. The nature of self-awareness is now a matter of serious discussion.

The work of Prof. Igor Aleksander, of the Electrical Engineering Department, Imperial College, London, involves exploring engineering definitions of intelligence, and asking which are the characteristics a system would need to exhibit for it to be termed intelligent, perhaps even termed self-aware. Engineering departments developing computers and robots now employ psychologists to help formulate problems in machine-perception to improve the ability of robots to recognise things. Engineers study consciousness and the working of the brain with technical applications in mind.

PHILOSOPHY, HISTORY & ENGINEERING

Philosophical matters need to be considered if engineering design is to be competently executed. Likewise, a new kind of historian is needed, able to analyse very recent developments and to serve a design team working on contemporary problems.

The life cycle of some recent technologies is a few years or months, and historians cannot restrict themselves to the events predating their birth or 30 years ago : too much has happened in the last five years which demands analysis, classification and interpretation. These historians will also need to be philosophers to understand the nature of the changes which they describe. Philosophers will also be needed in the education of engineers to prepare them for their duties in the next century. The links which are established between disciplines once remote from each other are connecting such subjects as evolutionary genetics ; engineering studies of the brain ; neuroscience ; intelligence ; information theory ; and bioengineering. There is attempted integration with modern theories of matter, and theories of how material ensembles can evolve and program themselves.

A comprehensive and powerful system for interpreting mankind in scientific, physical terms is being created with capacities much greater than the models used by Philosophical Materialism in the past. Indeed, Philosophical Materialism is undergoing a great revival and the cultural consequences will be considerable, not least for revealed, supernaturalist religions, and traditional systems of morality, ethics and socio-political organisation.

The philosophical discourses concerning the nature of things, and the possibility that man is a machine, associated with Enlightenment savants like La Mettrie, or D'Holbach, and which became a continual theme in 19th and 20th centuries humanism and freethought, remain vital issues. Today, they are kept alive as much by engineering developments as by developments in physics and biotechnology. Consequently, they are issues being spread throughout society.

Modern technology, in all its forms, is forcing a far wider section of society to consider such questions than ever did before. In the next century developments in artificial intelligence, artificial life, bio-engineering interfacing, and virtual reality — to name but a few areas where considerable development can be anticipated — will compel many to ask the question " What are we ? ".

Many of the author's students accept that human beings are very advanced, self-programming biological machines which have evolved and, using their own engineering abilities, will plan further, increasingly rapid development of the species with increased lifespan, and enhanced intelligence. That so many take this for granted is one cultural consequence of the changing nature of engineering.

DEVELOPING THE HISTORY OF ENGINEERING

It behoves all historians of engineering to ask what will be the impact of these developments on their discipline. The author believes that a radical and comprehensive development of those activities labelled " history of engineering " is required if an adequate account of the evolution of engineering from origins to present is to be compiled. History of engineering as it stands now is defective, incomplete and misleading, through a neglect of the methods and the conceptual apparatus of engineering, and a persistent failure to analyse engineering artefacts as embodiments of ideas. An overconcentration on equipmental form has resulted in too many detailed chronologies being passed off as history.

The general failure of professional historians and philosophers to understand engineering means that the few (very few) attempts from their ranks to tackle the problem fall short of what is required. History of engineering must start with engineering. No-one who does not understand engineering method, and who cannot solve engineering problems can write the mature engineering history which contemporary engineering requires. They might well write sound industrial archaeology, or conduct social studies of engineering communities, or compile histories of the consequences of engineering innovation on culture, but it is most unlikely that they will be able to do justice to history of engineering. The author suggests that engineers should take up the challenge themselves. Engineers, who have mastered the methods of conducting historical enquiry as a professional historian understands them, should direct their own program, working closely with the professional institutions and engineering centres in universities. They will need to work with historians and philosophers, and become professional in history and philosophy, but they must retain control of their own program and not assume that the people best suited to direct history and philosophy of engineering are historians or philosophers from outside. They are not. In the main, historians and philosophers share too many misconceptions about engineering to be of much help, though there are noteworthy exceptions. Besides, the coming together of many developments, rooted in engineering change, is so revolutionary that it is best to make a fresh start. Too many disciplines which matured before the recent wave of radical innovations lack the insights, and flexibility of mind, to respond to the challenge quickly enough. They are too conservative, and rooted in a culture which scientific-technological change is rendering meaningless.

The crisis in internalist history and philosophy of engineering is exacerbated because the tools for investigating it have to be forged rapidly in the face of rapid transformation of the nature of the object of study. Rather than try to use the methods of history of engineering as it is now, or to borrow from other departments of history, it is better to start anew, independent of those organisations (mainly learned societies and clubs) which at present claim expertise in history and philosophy of engineering.

The first thing to do is to recognise what modern engineering is, and to devise methods, concepts and constructs for defining the nature of engineering as represented by the most recent exemplars. These exemplars will need to be distinguished from those components and systems which originated in different eras. The several lines of evolution by which the different classes of engineering have evolved require tracing, with as much attention being given to conceptual apparatus and formative ideas as to equipmental form. An analytical, conceptualised history covering the developments of several millennia will be needed to chart how the nature of engineering has changed since activities worthy of the name " engineering " first appear in history. The concepts, theories, components and systems needed to understand modern engineering as represented by artificial intelligence, nanoengineering, virtual reality systems, or advanced robotics, indicate that engineering cannot be studied without recognising it as an expression of mind. To understand it requires an analytical, conceptualised history of change, and a philosophy to classify the nature of the changes.

Once this is grasped, the next step is to construct an adequate history and philosophy of engineering which traces the evolution of the many strands of engineering development through their major stages from the origins of human thought, or from an earlier period before the emergence of philosophy and articulate language. Recognising what engineering has become today will indicate which activities in the distant past will need to be studied, though every care must be taken to avoid the errors of " writing history backwards " and projecting modern assessments into previous ages. But much of the current attitudes to engineering are distorted by narrow 19th century prejudices and a tendency of amateur writers to define the nature of engineering through narrative descriptions of the changing forms of industrial equipment. So many learned societies, and history societies are dominated by this erroneous outlook that they are of little use in meeting the challenge modern engineering presents to the historian and philosopher.

Modern engineering has changed our perspective on engineering development, all the way back to the stone axe, and the first correspondence between artefact, mind and symbolic representation. Many of the traditional history of engineering societies are now antiquities, preserving outdated notions of history of engineering formed by amateur industrial archaeologists : they cannot meet the challenge of the recent insights into the nature of engineering which have arisen out of recent innovations.

The required history and philosophy will be internalist : only internalism can provide the engineering insight which alone can lead to adequate methods for analysing the innovations in different periods, so that they can be set in historical perspective. The studies will need to accommodate an expanded innovation analysis which accommodates an endless stream of innovations in the standards setting strategic areas which drive engineering forward. This is why the historians and philosophers required for the task are more likely to be drawn from the ranks of engineers with an enthusiasm for engineering progress than from the schools of history and philosophy as they are organised at present. This is not to say that the engineers can do it all alone, and work in isolation. As remarked above, engineering change is establishing links between engineering and disciplines such as psychology, neuroscience, and cognitive studies. Philosophers and historians working in these disciplines ought to be welcomed as collaborators. Those willing to work with engineers are recognisable by the work they do which carries them naturally towards engineering, just as the engineers (working for instance in AI, or VR) are naturally brought into contact with historians, philosophers, and other professionals in the course of their engineering work. But engineers must make a start themselves, in their own camp, on their own initiative, and prove themselves as engineer-historians and engineer-philosophers, before reaching out to the historians and philosophers developing within other strategic disciplines. They must not begin by turning to the historians and philosophers in the traditional, non-engineering societies.

Once the above steps have been taken, links can be established with historians and philosophers willing to develop history and philosophy in a modern-minded spirit : they will look back to understand the past, to turn towards the present, and to prepare for the future. Their history and philosophy will solve problems as well as deepen insight and extend knowledge. Much of existing history of engineering, narrative chronology, and industrial archaeology will be put to good use for the first time by relating them to the history of ideas, without which they remain pointless accumulations of technical facts. Much history of engineering, despite hard work, pedestrian diligence, enthusiasm, and persistence remains a monument to futility because it serves no enlightening end. But if the objects studied by the more competent chronologers of the " amateur " period are set into the evolving background of ideas which created them, and are provided with the philosophical analysis which they require, then much of this sterile history will bear fruit, in some cases long after the death of the compilers.

The new history will liberate historians from an obsession with things and perhaps revolutionise the policy behind accumulating collections in museums. It will begin with history, theories of history, philosophy, and the ever-changing conceptual apparatus by which engineering artefacts are defined, designed, recognised and analysed. When a history of engineering going back to the origins of systematic thinking has been devised, a rational policy for classifying objects, and developing collections may be formulated. Increasingly, within history of engineering, the emphasis will shift towards compiling accurate records of the abstract, the imaginary, and the theoretical attributes of engineering rather than the tangible or concrete equipment which can be seen and touched, because it is these entities which are the vital elements in the new engineering systems. To an engineer working in AI, VR, or cognitive studies, the evolution of mankind's use of language and symbols ; the emergence of perception ; the comparisons of human and animal intelligence set in evolutionary perspective, are likely to be much more important in the development of engineering, and to the understanding of its history, than those subjects found in most history of engineering journals. The emergence of articulate language ; the evolution of thought in relation to the bicameral brain ; and the history of studies of insects' intelligence are likely to become as much part of the history of engineering in the 21st century. as are studies of the Newcomen engine in the history of engineering of the 20th century. Investigators of AI and AL find theories of evolution, mind and genetics as relevant as strength of materials, perhaps more so : therefore the historians of these new departments of engineering will have to study them also. Granted the very rapid pace of innovation, with radical changes taking place within five years, these historians are needed today.

THE ELECTRONIC MUSEUM

Focusing the attention on the abstract, on the conceptual, and on cognitive models within the context of engineering, will be encouraged by the creation of electronic mail, electronic archives, and electronic museums. When there is available a world wide network, able to exchange accurate images of artefacts, and providing access to archives, it should be possible to provide an integrated display of an image set in its philosophical, conceptual and theoretical framework.

Visual representation, pictures, engineering drawings, network diagrams, mathematical analysis, experimental data, etc., could be called up as part of an information package dedicated to that particular item. Computer replication and simulation may make it possible, in the near future, for the historian to explore electronically created replicas of any kind of device or system collected in the electronic museum. Designs could be analysed, inspected, taken apart, operated, and understood in a way simply not possible with actual hardware replicas. The analysis, using computer replication, could be extended to bio-technical ensembles, such as the microprocessor implanted into a rat's brain ; or into simulations of the small scale interiors of nanoengineering systems, or into the depths of " Colossus ". Granted the anticipated improvements to electronics information systems of all kinds, there does not seem to be any reason why replicas and models of engineering systems from any period should not be created and stored within electronics museums. This will help concentrate attention on history of engineering as " knowledge and ideas ", relying on sound analysis, drawing on good records and archives. The global electronics-mail museum, using virtual reality systems, would present images of artefacts in such a way that the connection between visual representation of an item and the associated conceptual apparatus and philosophical framework was made evident. This would liberate museum displays from the misleading domination of the perceived object, and render unnecessary the conservation of buildings and machines which had been accurately recorded using electronic means.

The stress should switch from the preservation of objects to the preservation of information. If scholarship and enlightenment are the objects of historical study, much of what is today termed history of engineering is due for a change in every department.

The learned societies will either carry through extensive and painful trans-formations, and belatedly professionalise themselves, or they will fossilise within an older tradition of amateur history, quite inadequate in the 21st century, and survive as clubs which organise visits and meetings representative of " history as it used to be ". They will become part of the " heritage history " movement. The author believes that a mature history of engineering can only be created by the engineering profession itself. When this has been done, fruit-ful links can be established with historians and philosophers working within bodies like IUHPS, ICOHTEC, BSPS, SHOT, etc., but not before.

LES INGÉNIEURS DE LA RENAISSANCE ET LEURS MANUSCRITS ET TRAITÉS ILLUSTRÉS

Eberhard KNOBLOCH

INTRODUCTION

" Un ingénieur n'est pas un architecte commun, mais un architecte de fortifications et de guerre qui a un esprit perçant pour jalonner une fortification, un endroit ou un retranchement… Il doit être parfaitement expérimenté dans l'arithmétique, il doit avoir aussi de l'expérience pratiquée dans la géométrie, les mathématiques et la mécanique, il doit maîtriser au moins théoriquement l'artillerie et tout ce qui concerne les maîtres bombardiers "[1].

Ainsi l'arithméticien et ingénieur de fortifications d'Ulm Johannes Faulhaber décrit-il les talents, la formation préparatoire et la compétence d'un ingénieur en 1633. Sa description nous révèle trois traits caractéristiques d'un tel homme : 1. l'identité avec l'architecte, 2. l'importance des mathématiques, 3. la relation étroite avec l'art militaire. Cette caractérisation doit être considérée et complétée de deux aspects, si nous envisageons les ingénieurs des deux siècles précédents, à savoir de la technique civile et de l'activité artiste, d'autant plus que les mêmes critères d'évaluation s'appliquent à l'art et au génie.

La notion d'invention était une notion-clé de la théorie de l'art de la Renaissance, de cette époque qui adora la créativité de l'individu. Fertilité en inventions et sagacité étaient les propriétés dont les ingénieurs de la Renaissance se flattaient. Francesco di Giorgio Martini parlait même d'une procédure infinie, pourvu qu'il voulût décrire tous les instruments inventés[2]. La remarque nous

1. J. Faulhaber, *Anderer Theil der Ingenieurs Schul*, Ulm, 1633, 3 ; G. Zweckbronner, " Rechenmeister, Ingenieur und Bürger zu Ulm - J. Faulhaber (1580-1635) in seiner Zeit ", *Technikgeschichte*, 47 (1980), 114-132, voir 114.

2. F. di Giorgio Martini, *Trattati di architettura ingegneria e arte militare*, vol. 1, a cura di Corrado Maltese, trascrizione di Livia Maltese Degrassi, Milan, 1967, 505 (2 vols) ; M. Kemp, " From " Mimesis " to " Fantasia " : the Quattrocento vocabulary of creation, inspiration and genius in the visual arts ", *Viator*, 8 (1977), 347-399, voir 396.

rappelle Vitruve qui avait loué les nombreuses admirables inventions d'Archi-mède et son esprit créateur infini[3]. Par conséquent, je voudrais mentionner les cinq points de vue suivants :

1. Architecte et Ingénieur.
2. Machine et Auteur.
3. Les mathématiques et la technique.
4. Les descendants d'Archimède.
5. Fantaisie et utopie.

Conclusion : le triomphe de l'esprit.

ARCHITECTE ET INGÉNIEUR

Filippo Brunelleschi (1377-1446) s'élève au-dessus des nombreux techni-ciens et ingénieurs italiens doués, et ceci d'une manière révélatrice. Sa renon-ciation complète à des notes écrites qui l'auraient rendu auteur au sens de l'humanisme, l'instruction donnée à son élève et ami Mariano Taccola — son cadet de quelques années seulement — de ne pas rendre public ses inventions rappellent l'attitude des architectes médiévaux.

Ses activités les plus diverses telles que horloger et orfèvre, peintre et tech-nicien, en tant qu'organisateur de fêtes civiles et religieuses, architecte et ingé-nieur militaire le font apparaître comme un artiste-ingénieur typique de la Renaissance. Au XVIᵉ et même au XVIIᵉ siècle encore, les auteurs de livres de machines indiquèrent sur les frontispices de leurs oeuvres qu'ils étaient en même temps des architectes et des ingénieurs : Salomon de Caus, Giovanni Branca, Georg Böckler peuvent servir d'exemples[4]. De même que ses succes-seurs Taccola avec Mariano Sozzini, Francesco di Giorgio Martini avec Ubal-dini, Léonard de Vinci avec Luca Pacioli et Giorgio Valla[5], il a cherché et trouvé des contacts amicaux avec des humanistes importants tels que Paolo dal Pozzo Toscanelli, ce qui caractérise le nouveau rang social de ces ingénieurs. En introduisant la perspective dans la peinture, il enseignait la connaissance de la réalité à l'aide des mathématiques.

Face à des projets techniques ambitieux et risqués, des échecs ne pouvaient manquer de se produire. Brunelleschi aussi devait faire cette expérience. Son essai d'inonder Lucques par la rivière Serchio qui était assiégée par les Floren-

---

3. Vitruve, *Les Dix livres d'Architecture*, IX, 9.

4. S. de Caus, *Von Gewaltsamen bewegungen Beschreibung etlicher, so wol nützlichen alß lustigen Machiner beneben Underschiedlichen abriessen etlicher Höllen oder Grotten und lust Brunnen*, Francfort-sur-le-Main, 1615 ; G. Branca, *Le machine volume nuovo e di molto artificio da fare effetti maravigliosi tanto Spiritali quanto di Animale operatione arichito di bellissime figure con le dichiarationi a ciascuna di esse in lingua volgare et latina*, Rome, 1629 ; G. Andreas Böckler, *Theatrum machinarum novum, Schauplatz Der Mechanischen Künsten von Mühl- und Wasserwercken*, Nuremberg, 1661.

5. P. Galluzzi, *Les ingénieurs de la Renaissance de Brunelleschi à Léonard de Vinci*, Paris, 1995.

tins en 1428 échoua : les eaux de la rivière détournée se jetèrent sur les assié-
geants.

La prudence dont l'histoire n'est pas écrite jusqu'à maintenant était ainsi
justifiée devant les idées audacieuses des ingénieurs, quand les autorités de
Florence devaient décider de la manière dont il faudrait construire la coupole
de la cathédrale de Florence. La technique des voûtes était le talon d'Achille
de beaucoup d'architectes. Nous y reviendrons. Brunelleschi concurrençait
Lorenzo Ghiberti comme en 1401. Mais cette fois-ci, ce fut lui que l'on char-
gea de réaliser les travaux. Le problème technique résidait dans les dimensions
énormes du bâtiment : On voulait construire une coupole sur l'octogone, dont
l'ouverture était de 55 mètres et cela environ à 54 mètres au-dessus du sol
(Illustrations 1 et 2). Brunelleschi — c'est bien connu grâce aux travaux de
Paolo Galluzzi — proposa une construction sans l'emploi du centre en bois
pour soutenir la maçonnerie en cours d'exécution. Il assura qu'une telle technique était possible grâce à la forme raide elliptique de la coupole. Le succès
lui donna raison.

ILLUSTRATION 1.

La cathédrale de Florence : B. Evers, *op. cit.* (voir note 9), 15.

ILLUSTRATION 2.

La lanterne de la coupole de la cathédrale de Florence :
Galuzzi, *op. cit.* (voir note 5), 13.

Il est hors de doute qu'il ne disposait pas d'une théorie mathématique qui lui fournit la figure géométrique, statiquement stable. Seul le mathématicien David Gregory prouva en 1697[6] que la ligne de soutien théoriquement correcte de voûtes était une chaînette inversée. Les fissures dans la coupole florentine étaient inévitables, parce que Brunelleschi— pour parler mathématiquement — n'avait pas utilisé, n'avait pas pu utiliser de solution minimale. Les semblables problèmes de construction de la cathédrale St. Pierre sont bien connus[7]. Mais une analyse mathématique du problème et des considérations mécaniques, dont nous ne savons rien faute de documents et dont l'existence ne peut cependant pas être contestée, lui rendirent possible une solution et firent de lui un pionnier du génie basé sur les mathématiques.

Pour mener sa tâche à bonne fin, Brunelleschi dut faire les plans de beaucoup de machines de construction, particulièrement de grues, inventer des

6. J. Heyman, " The theory of structures ", I. Grattan-Guinness (éd.), *Companion Encyclopedia of the History and Philosophy of the Mathematical Sciences*, London, New York, 1994, 1034-1043, voir 1040.

7. I. Szabó, *Geschichte der mechanischen Prinzipien und ihrer wichtigsten Anwendungen*, Bâle, Stuttgart, 1976, tables VIII-XI. H. Bredekamp, *Sankt Peter in Rom und das Prinzip der produktiven Zerstörung, Bau und Abbau von Bramante bis Bernini*, Berlin, 2000.

solutions particulières de la technique du bâtiment et disposer d'une excellente organisation du chantier. Nombre de ses machines furent dessinées par ses successeurs tels que Taccola, Léonard ou Buonaccorso Ghiberti, par exemple la grue suivante par Léonard[8] :

ILLUSTRATION 3.

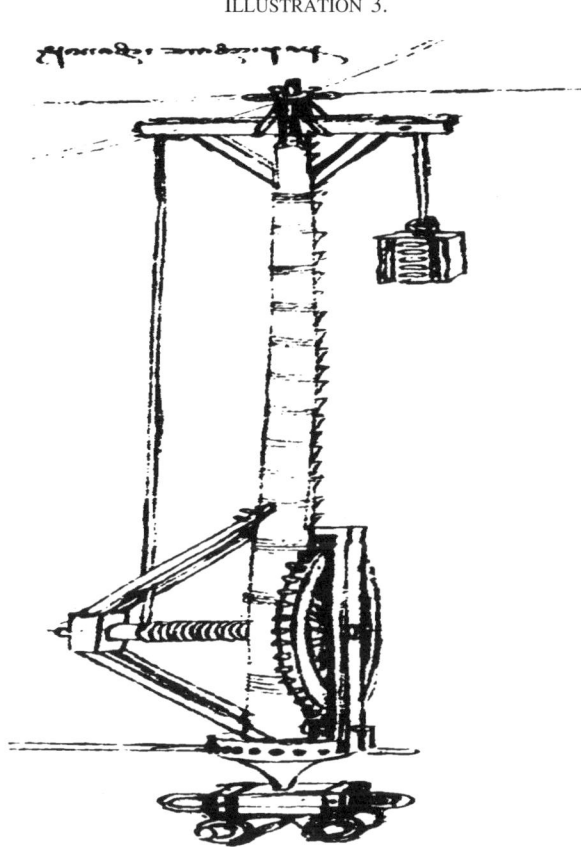

Léonard de Vinci, Un élévateur de Brunelleschi :
Codex Atlanticus, f. 138r.

On la retrouve sur une image magnifique, sur la fameuse *spaliera*, le dossier au sens propre, la table de Sarasota en Floride, qui est un des témoignages les plus expressifs et les plus remplis de vivacité en ce qui concerne la technique de construction à la Renaissance[9]. Par conséquent, il nous faut la considérer plus précisément.

8. P. Galluzzi, *op. cit.* (voir note 5), 99-116, voir 116.
9. B. Evers (éd.), *Architekturmodelle der Renaissance, Die Harmonie des Bauens von Alberti bis Michelangelo*, Munich, New York, 1995, 208 s.

ILLUSTRATION 4.

Piero di Cosimo, La construction d'un palais :
B. Evers, *op. cit.* (voir note 9), 209.

De telles tables ont pour thème des sujets mythologiques et allégoriques qui embellissaient les palais de cette époque. On s'accorde à attribuer ce tableau à la fin de la période de créativité 1515-1520 du peintre florentin Piero di Cosimo (1462-1521).

Le motif principal est la construction d'un palais consistant en deux parties, dont la représentation comprend tous les détails de ce qui se passe sur un chantier : d'après le théoricien antique de l'architecture Vitruve et son fameux rénovateur humaniste Alberti, le chantier est montré comme le vrai lieu d'enseignement et d'apprentissage. Que voulait dire Goethe sinon cela, lorsqu'il écrivit dans ses cahiers sur la morphologie en 1822 : " A vrai dire, même dans les sciences, on ne peut rien savoir. Il faut toujours que ce soit fait "[10].

Tandis que le regard est captivé par le palais double, dessiné en perspective centrale, il saisit en même temps les mulets avec les matériaux de construction, les charpentiers avec les haches, les ouvriers des transports avec l'attelage des bœufs, les tailleurs de pierres avec le marteau et le ciseau, mais aussi les trois ouvriers qui tirent un chariot avec une statue, les quatre hommes qui portent une poutre, les deux menuisiers avec une scie, l'architecte qui explique au moyen d'une équerre la construction du fût d'une colonne à un jeune homme, les gens qui érigent une statue sur le rebord du palais au moyen d'une grue. Une telle grue est décrite par Vitruve, fut utilisée par Brunelleschi et dessinée par Léonard, comme nous l'avons vu. L'idéal humaniste de la symétrie caractérise toute la composition du tableau.

10. J.W. von Goethe, " Maximen und Reflexionen, Aus den Heften zur Morphologie ", dans J.W.v. Goethe, *Poetische Werke, vollständige Ausgabe*, vol. 2, Stuttgart, 1950, 705-711, 711 (maxime 415).

La peinture contient en même temps beaucoup de détails allégoriques complémentaires qui méritent d'être considérés un peu plus précisément : la mère qui montre le chantier à son enfant, les enfants se balançant, les jeunes gens suspendus ou ceux qui jouent dans la cour, le cavalier sur le cheval blanc, le concours hippique et les deux cavaliers, apparemment un père avec son fils au premier plan. On les a interprétés — à mon avis avec raison — comme Dédale et Icare. En fait, toutes ces scènes symbolisent l'aspect de jeu dans diverses activités — différentes occupations, enseignement, apprentissage, etc. — et lient l'esprit du jeu à l'habilité mécanique. Cet aspect répond en même temps à la fertilité en inventions[11]. En conséquence, il se trouve constamment souligné dans les manuscrits techniques illustrés de la Renaissance. Dédale est sa personnification. Prenons à titre d'exemple le " Manuel du moyen-âge " qui fut rédigé vers 1480, à proprement parler un livre élargi pour les maîtres bombardiers. Il contient beaucoup de scènes de bain (Kyeser[12] fait la même chose), de chasse, de tournoi et d'amour dans des jardins avec une roue hydraulique[13].

ILLUSTRATION 5.

Manuel du Moyen-Age, Scène de chasse : A. Essenwein, *op. cit.* (voir note 13), f. 12a.

11. H. Bredekamp, *Antikensehnsucht und Maschinenglauben, Die Geschichte der Kunstkammer und die Zukunft der Kunstgeschichte*, Berlin, 1993 (2e éd. 2000), 66 s. " Schöpfung als Spiel ".

12. C. Kyeser, *Bellifortis*, dans Götz Quarg (éd.), Düsseldorf, 1967, 2 vols.

13. A. Essenwein (éd.), *Mittelalterliches Hausbuch, Bilderhandschrift des 15. Jahrhunderts mit vollständigem Text und facsimilierten Abbildungen*, Francfort-sur-le-Main, 1887 ; E. Knobloch, " Übergang zur Renaissance - Deutsche Tradition ", in U. Lindgren (éd.), *Europäische Technik im Mittelalter, Tradition und Innovation*, 2e éd., Berlin, 1997, 569-582, voir 573.

ILLUSTRATION 6.

Manuel du Moyen-Age, Une scène d'amour avec une roue hydraulique :
A. Essenwein, *op. cit.* (voir note 13), f. 24b - 25b.

Le père de la métallurgie Vannoccio Biringuccio n'oublie pas d'insérer un chapitre sur les feux d'artifice pour un spectacle et sur le feu de l'amour dans son oeuvre sur la pyrotechnie[14].

MACHINE ET AUTEUR

A la différence de Brunelleschi une production de traités techniques illustrés et de plus en plus riches se développait encore de son vivant. Le traité sur l'art des ingénieurs était une création de la Renaissance. Les auteurs tiraient leur origine d'abord d'Italie et d'Allemagne, au XVIᵉ siècle aussi d'Espagne, de France et de la Belgique. On écrivait souvent dans les langues nationales et non en latin. On se rattachait aux techniciens et aux auteurs antiques de livres militaires comme Archimède, Vitruve, Pline, Frontin, Végèce, dont les oeuvres furent imprimées peu après l'invention de l'impression typographique, ainsi qu'aux auteurs arabes, comme Giovanni Fontana. Mais on les surpassait en même temps d'une manière créatrice, ce que Kyeser, ce que Fontana dirent expressément[15].

14. V. Biringuccio, *De la pirotechnia libri X*, Venise, 1540 (rééd. Milan, 1977), 159r-160r, 166v-168r.

15. C. Kyeser, *op. cit.* (voir note 12), vol. 2, 4 ; E. Battisti et G. Saccaro Battisti (éds), *Le macchine cifrate di Giovanni Fontana. Con la riproduzione del cod. Icon. 242 della Bayerische Staatsbibliothek di Monaco di Baviera e la decrittazione di esso e del Cod. Lat. Nouv. Acq. 635 della Bibliothèque nationale di Parigi*, Milan, 1984, 69.

La "renaissance des machines" allait de pair avec l'émancipation et la mathématisation de la peinture, avec l'usage de descriptions visuelles, avec l'usage de la langue des images.

La notion de machine changeait sa signification avec le temps et doit être mise dans un large sens du mot. Fontana l'évitait complètement en utilisant les expressions *ingenium* ou *opus*, ouvrage. Le deuxième mot nous rappelle évidemment un contexte religieux qu'on retrouve dans les livres de machines au XVIe et XVIIe siècles.

On était convaincu que la description textuelle ne suffisait pas pour comprendre des mécanismes techniques tandis que l'image d'une machine n'avait pas toujours besoin d'un texte explicatif. Taccola[16] disait souvent : le dessin parle de lui-même. Dans ce cas, la langue des images a remplacé la langue textuelle. C'est dans ce contexte culturel qu'il faut voir les livres du XVIe siècle intitulés *Théâtres des machines*. C'est le siècle des descriptions cosmographiques du monde, c'est le siècle de la renaissance de la métaphore "théâtre du monde"[17] créée par Platon et utilisée par John of Salisbury au XIIe siècle. Les théâtres des machines forment une partie du théâtre du monde. L'homme est un acteur en tant qu'ingénieur.

Alberti, il est vrai, publia son oeuvre sur l'architecture orientée vers Vitruve sans aucune illustration. Mais elle était adressée à des protecteurs et des mécènes, et non pas à des architectes. Ses successeurs et les traducteurs de Vitruve ne renoncèrent pas à ce moyen d'illustration. Ainsi ce n'était sans doute pas dû au hasard que le *Manuel du moyen-âge* commence par un texte latin qui soulignait l'importance cruciale d'images pour la mnémotechnique. La compréhension et la mémoire dépendent d'images, de dessins. L'ingénieur devint auteur au sens de l'humanisme. Son ouvrage immortalisa le patron auquel il fut dédié. Il faut cependant ajouter qu'il y avait des auteurs importants et influents de ce genre qui n'étaient pas des ingénieurs de métier, mais des médecins comme l'Allemand Conrad Kyeser et l'Italien Giovanni Fontana ou des humanistes comme Roberto Valturio qui enseigna rhétorique et la poésie à Bologne et au XVIe siècle Georg Agricola.

Le livre de Valturio sur l'art militaire[18] était principalement un traité sur l'art militaire romain, enrichi de 95 gravures sur bois et ainsi le premier livre imprimé avec des illustrations techniques et scientifiques sans être conçu comme une instruction technique pour construire des appareils militaires.

16. M. Taccola, *De rebus militaribus*, Baden-Baden, éd. par E. Knobloch, 1984.

17. E.R. Curtius, *Europäische Literatur und lateinisches Mittelalter*, 6e éd., Bern, Munich, 1967, 150.

18. R. Valturio, *De re militari*, Vérone, 1472 ; B. Evers (éd.), *Architekturmodelle der Renaissance, Die Harmonie des Bauens von Alberti bis Michelangelo*, *op. cit.* (voir note 9), 189.

ILLUSTRATION 7.

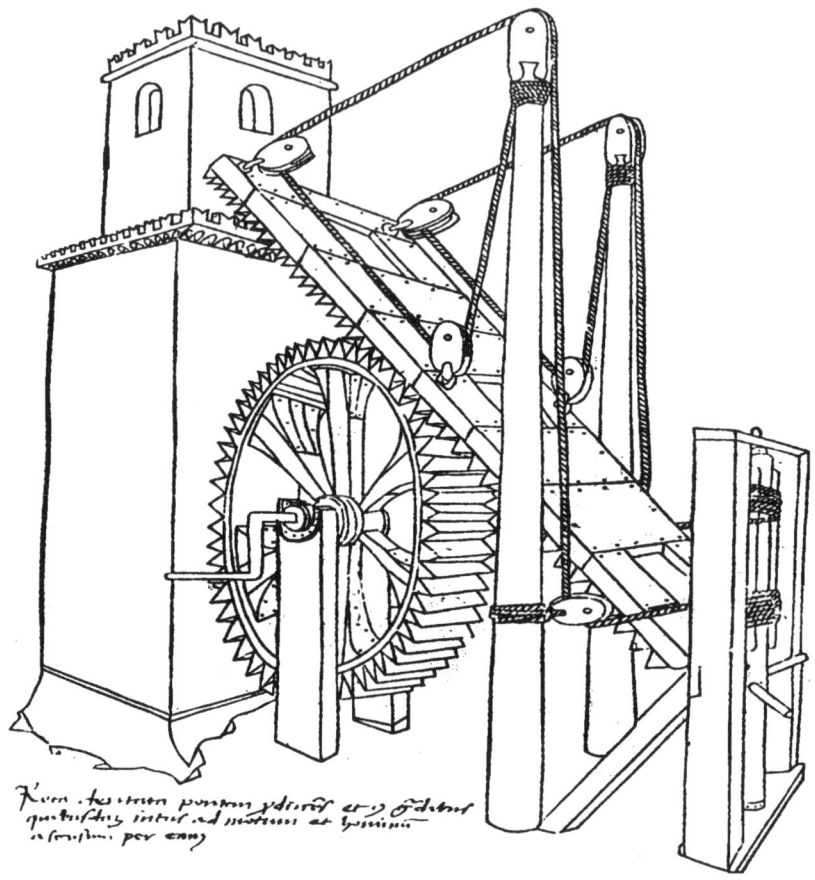

Roberto Valturio, Une échelle d'assaut :
B. Evers, *op. cit.* (voir note 9) 189.

Cette échelle d'assaut unit plusieurs principes de la mécanique élémentaire. On l'a copiée de même que d'autres machines pour l'édition parisienne de Végèce[19].

L'illustrateur siennois de livres, l'ingénieur militaire, l'hydraulicien et architecte Mariano Taccola qui écrivit en latin, peut être considéré comme un prototype de l'ingénieur de la Renaissance. Ses dessins techniques et esquisses étaient très souvent copiés et étudiés particulièrement par Francesco et Léonard. Ils traitent de tous les domaines du génie civil et militaire. Ils incluent

---

19. V. Marchis, " Macchine fra realtà e fantasia, L'orizzonte tecnico di Roberto Valturio ", *Le macchine di Valturio nei documenti dell'Archivio Storico Amma*, presentazione di S. Ricossa, Turin, 1988, 117-141, voir 122.

des techniques et machines antiques et modernes pour la guerre de siège et de défense, par conséquent des armes à feu aussi , des fortifications, des bateaux et des canots, des grues, des appareils élévateurs et des treuils, des moulins, des élévateurs d'eau. Considérons les quatre dessins à titre d'exemples :

ILLUSTRATION 8.

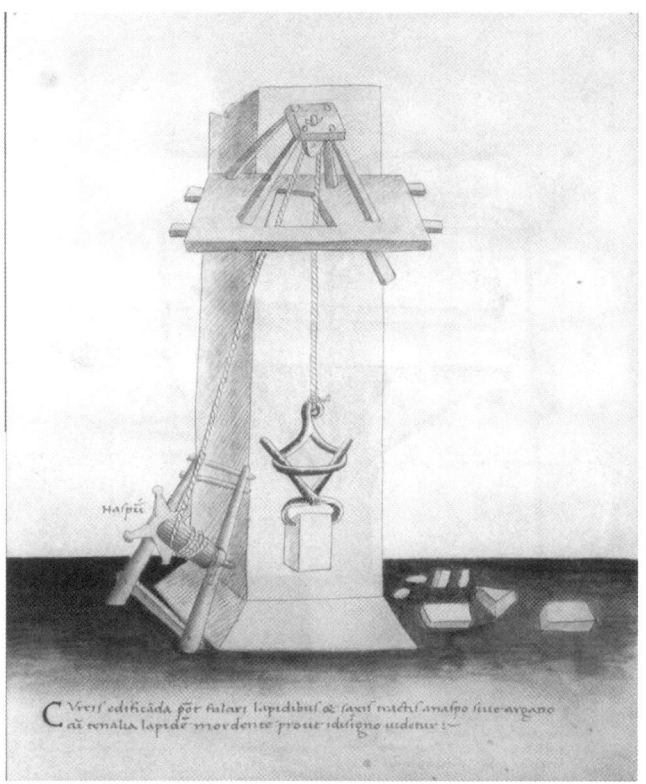

Mariano Taccola, Une grue de construction avec une pince automatique :
L'art de la guerre, Machines et stratagèmes de Taccola,
ingénieur de la Renaissance, présenté par E. Knobloch (Paris 1992), 119.

Un dispositif de levage avec une pince automatique ; Brunelleschi l'a utilisé pour la construction de la coupole florentine. Léonard l'a dessiné dans le Codex Madrid I[20].

Un tuyau d'aspiration lié à un tuyau d'écoulement qui fonctionnent d'après le principe d'un siphon. L'eau montant attirait l'intérêt de beaucoup d'auteurs.

20. L. da Vinci, *Codices Madrid*, Faksimileausgabe mit Transkription und Kommentar von Ladislao Reti und deutscher Übersetzung von Gustav Ineichen, Friedrich Klemm, Ludolf v. Mackensen, Reinhilt Richter, Francort-sur-le-Main, 1974, 5 vols ; cod. Madrid I, f. 22r.

On retrouve ce dispositif déjà chez Kyeser, plus tard chez Francesco[21], chez Valturio[22].

ILLUSTRATION 9.

Mariano Taccola, Un tuyau d'aspiration lié à un tuyau d'écoulement :
L'art de la guerre, op. cit., 139.

21. C. Kyeser, *op. cit.* (voir note 12), f. 61v ; B. Degenhart, A. Schmitt, *Corpus der italienischen Zeichnungen 1300-1450*, Teil II, *Venedig Addenda zu Süd- und Mittelitalien*, vol. 4, Katalog 717-719, *Mariano Taccola*, Berlin, 1982, 56-61.

22. V. Marchis, *op. cit.* (voir note 19), 127.

Un trébuchet, c'est-à-dire une catapulte à verge basculante. Déjà Kyeser le dessina, et encore Franz Helm au XVIe siècle[23]. (Illustration 10)

ILLUSTRATION 10.

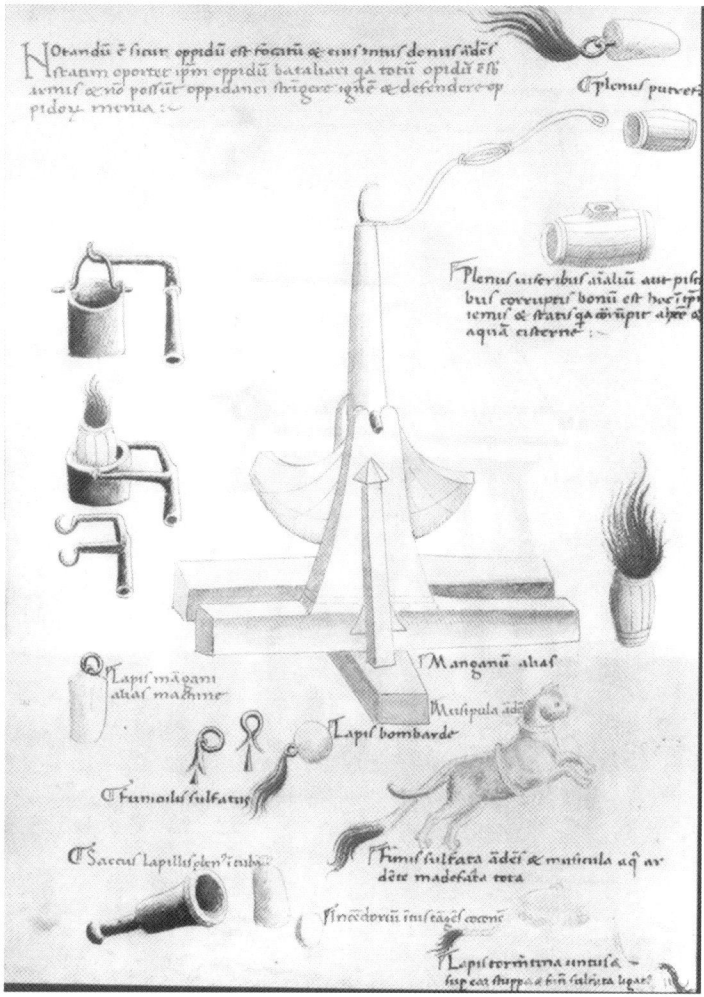

Mariano Taccola, Un trébuchet :
L'art de la guerre, op. cit., 42.

23. M. Taccola, *op. cit.* (voir note 16), 21 ; F. Helm, *Buch von den probierten Künsten*, Staatsbibliothek Preußischer Kulturbesitz Berlin, Haus 2, Ms. germ. quart. 1188, f. 169v ; Staats- und Universitätsbibliothek Göttingen, Philos. 65, f. 118v ; la reproduction des dessins est approuvée par la Staatsbibliothek Preußischer Kulturbesitz ; R. Leng, *Franz Helm und sein* Buch von den probierten Künsten, *Ein handschriftlich verbreitetes Büchsenmeisterbuch in der Zeit des frühen Buchdrucks*, Wiesbaden, 2001.

Des bombardes ou canons à pierres, dessinés comme arme se chargeant par la culasse et comme fusil à baguette. (Illustration 11)

ILLUSTRATION 11.

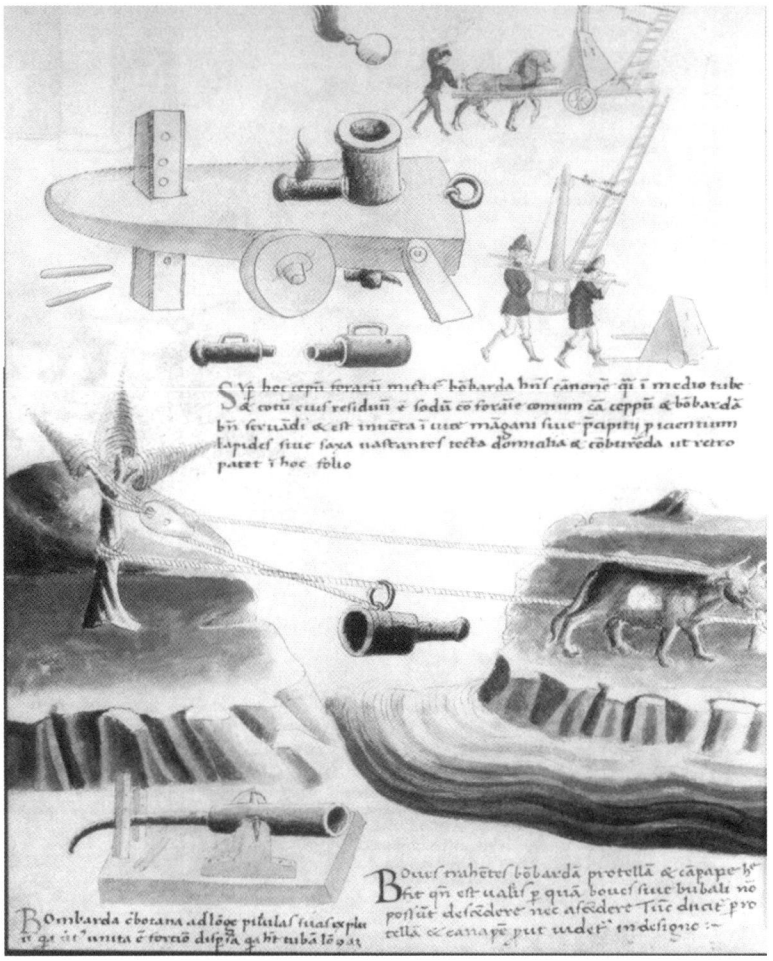

Mariano Taccola, Des bombardes :
L'art de la guerre, op. cit., 82.

De telles armes constituent le sujet principal des livres destinés aux maîtres bombardiers dont la plupart ne sont pas encore publiés jusqu'à maintenant. Le plus important du XVIe siècle, à savoir le *Livre des arts éprouvés* achevé vers 1535 par Franz Helm, ne fut publié qu'en 2001. On y trouve un trésor d'informations techniques qui furent transmises par de tels écrits, par exemple les dessins à l'encre suivants : (Illustrations 12-19).

ILLUSTRATION 12.

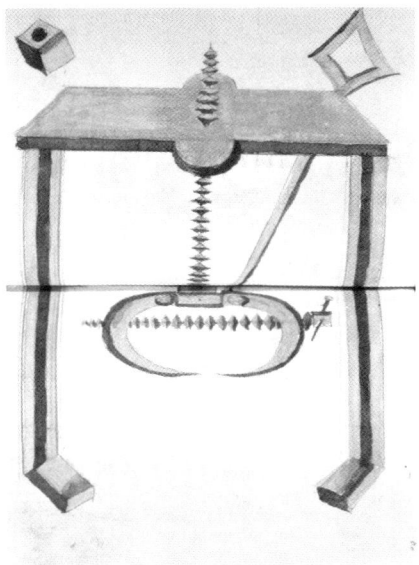

F. Helm, Un vérin pour élever une bombarde :
op. cit. (voir note 23), f. 59v - 60r.

ILLUSTRATION 13.

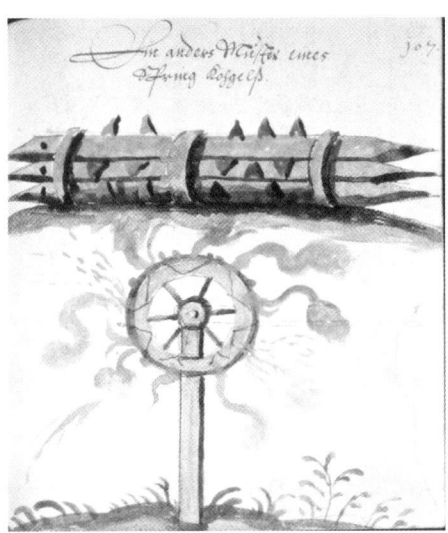

F. Helm, Un projectile et une roue de roquettes :
op. cit. (voir note 23), f. 107r.

ILLUSTRATION 14.

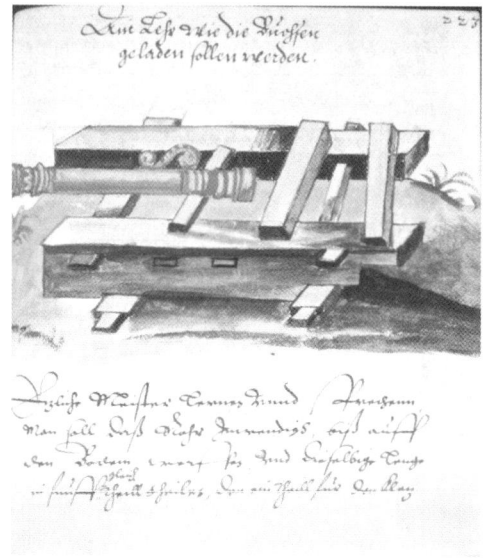

F. Helm, Une bombarde sur une base de bois :
op. cit. (voir note 23), f. 223r.

ILLUSTRATION 15.

F. Helm, Un trébuchet : *op. cit.* (voir note 23), f. 169v.

ILLUSTRATION 16.

F. Helm, Un bombardier qui charge un canon :
op. cit. (voir note 23), f. 226v.

ILLUSTRATION 17.

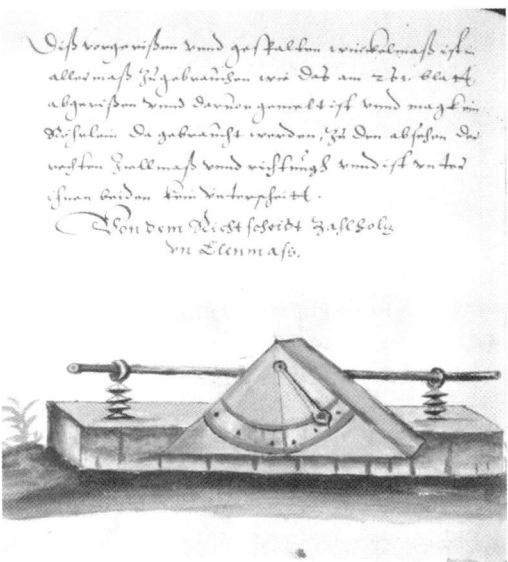

F. Helm, Un quadrant et une règle :
op. cit. (voir note 23), f. 265v.

ILLUSTRATION 18.

F. Helm, Un bombardier qui tire : *op. cit.* (voir note 23),
tableau pliant après f. 282.

ILLUSTRATION 19.

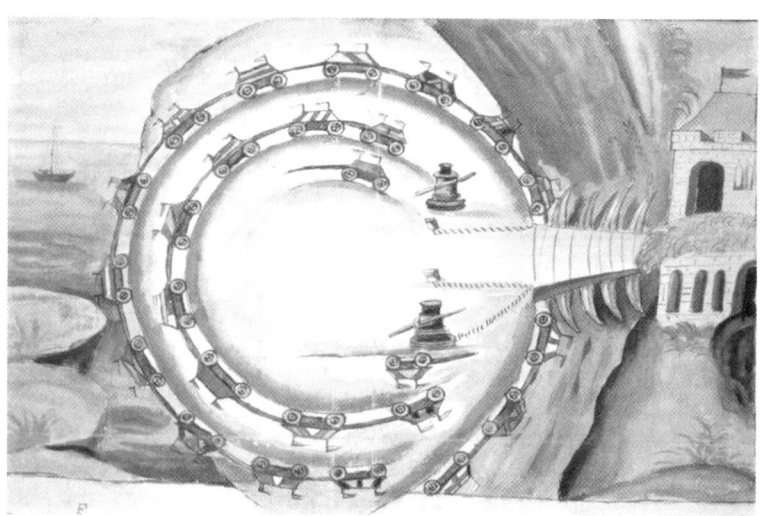

F. Helm, Une barricade hussite de chariots : *op. cit.* (voir note 23),
tableau pliant après f. 302.

1. Un vérin pour élever une bombarde ;

2. Un projectile et une roue de roquettes ;

3. Une bombarde sur une base de bois ;

4. Un trébuchet ;

5. Un bombardier qui charge un canon au moyen d'un bâton de charge ;

6. Un quadrant et une règle ;

7. Un bombardier qui tire ;

8. Une barricade hussite de chariots comme on l'a trouvée aussi dans le *Manuel du moyen-âge*.

LES MATHÉMATIQUES ET LA TECHNIQUE : THÉORIE, PRATIQUE, THÉORIE ET PRATIQUE

Théorie

Nul autre que le mathématicien le plus remarquable du XVe siècle, Jean Regiomontanus, mit en vedette la valeur des mathématiques pour tous les domaines de la vie et de la connaissance humaine. Son cours fait à Padoue en 1464 " Sur la dignité et l'utilité des mathématiques "[24] donne d'abord une classification des sciences mathématiques, au nombre desquelles il comptait la science des poids, des aqueducs et de la raison des vitesses en mouvements. Il évoqua l'astronome de Padoue Giovanni de'Dondi qui avait construit son horloge astronomique, sa machine du temps, pour citer Philippe de Mézières[25], environ cent ans plus tôt. Cette horloge liait la beauté à l'utilité, comme Regiomontanus le dit[26], et fut admirée par bon nombre de gens comme un miracle pour ainsi dire. (Nous reviendrons à cette relation entre technique et miracle un peu plus tard.) Deux choses méritent d'être mises en évidence :

1. De Dondi élabora un long traité avec des dessins sur sa construction. La langue des images était inévitable.

2. Pour réaliser cette machine compliquée, une vraie collaboration entre la science, les arts libéraux et la technologie, les arts mécaniques, fut nécessaire, collaboration qu'il n'y avait guère ailleurs[27].

Ainsi les développements de Regiomontanus se transformèrent-ils en une glorification enthousiaste des sciences mathématiques. Il expliqua qu'il passait sous silence l'activité des mécaniciens et des artisans. Que la géométrie les

24. J. Regiomontanus, *Oratio habita Patavii in praelectione Alfragani*, Nuremberg, 1537 ; rééd. dans J. Regiomontanus, *Opera Collectanea*, in F. Schmeidler (éd.), Osnabrück, 1972, 43-53.

25. C.M. Cipolla, *Die gezählte Zeit - Wie die mechanische Uhr das Leben veränderte*, Berlin, 1997, 48.

26. J. Regiomontanus, *op. cit.* (voir note 24), 47.

27. J. Gimpel, *La révolution industrielle du Moyen Age*, Paris, 1975, 156.

guiderait le mieux, si ces gens-là suivaient ses prescriptions dans la construction des bâtiments, dans le déversement des eaux ou le déplacement des poids. Qu'il pouvait rappeler combien de fois les nouvelles voûtes d'églises s'étaient écroulées à cause de l'insouciance complète des architectes — *admodum ignavia architectorum* — qui utilisaient des figures inadéquates. Que récemment quelques moines avaient été écrasés à Venise quand une tour tomba à la renverse. On a attribué ce malheur à l'artisan illettré, *ruditati opificis*, qui essayait de l'élever.

Cinquante ans plus tard, Luca Pacioli, l'ami de Léonard de Vinci, a continué la description idéalisante de Regiomontanus dans l'optique humaniste. Mais sa conclusion diffère.

En 1509, Pacioli met en avant dans sa Divine proportion, que les disciplines mathématiques sont le fondement et l'échelle pour parvenir à la connaissance de chaque autre science[28]. Déjà Archimède, le géomètre ingénieux et l'architecte le plus digne, *ingegnoso geometra et dignissimo architetto*, a défendu sa patrie au moyen de machines et d'instruments, *con ingegni e instrumenti*, c'est-à-dire au moyen des mathématiques. Toutes les artilleries, instruments et machines militaires sont construits d'après les disciplines mathématiques. De cette manière la louange des mathématiques de Pacioli devient à l'improviste une louange des ingénieurs à la différence de Regiomontanus : les mathématiques confèrent sa réputation aux ingénieurs, elles les ennoblissent.

Les vieux romains étaient victorieux parce qu'il prenaient diligemment soin des ingénieurs, *ingegneri*. Aucune bonne armée envoyée à une attaque ou à la défense ne peut se nommer bien munie de tout si elle ne dispose pas d'ingénieurs et d'un constructeur particulier de machines nouvelles, *novo machinatore particular ordinato*.

Pratique

La sévère critique par le théoricien humaniste de leur ignorance mathématique et par conséquent de leur incapacité au demeurant très dangereuse trouve écho dans les plaintes du praticien, ingénieur et architecte Francesco di Giorgio Martini. Francesco rédigea son traité sur l'architecture civile et militaire vers 1475, c'est-à-dire environ dix ans après le discours de Regiomontanus. Le dernier chapitre traite des machines à mouvoir des poids, des aqueducs, des moulins. Francesco s'y plaint sévèrement des plagiaires qui usurpent les inventions d'autres personnes[29]. Il dit que ce vice est répandu à l'excès à son époque et ceci particulièrement parmi les architectes, *li quali sono quasi tutti omini igno-*

---

28. L. Pacioli, *Divina proportione, Die Lehre vom Goldenen Schnitt*, Nach der Venezianischen Ausgabe vom Jahre 1509 neu hrsg., übersetzt und erläutert von C. Winterberg, Vienne, 1889, 185.

29. L. Reti, " Francesco di Giorgio Martini's Treatise on Engineering and its Plagiarists ", *Technology and culture*, 4 (1963), 287-298, voir 291 s.

ranti et inesperti[30], " qui presque tous sont des hommes ignorants et inexpéri-mentés, ce que l'on peut facilement comprendre face à leurs oeuvres ".

Le dépit personnel pourtant très justifié de Francesco le fit aller plus loin que son but. A vrai dire il n'est pas le seul à cet égard. Michelangelo appela ses collègues même moutons et boeufs[31].

Les remarques critiques montrent quand même, que la rétrospection histori-que mène aisément à une image transfigurée au regard des personnes célèbres comme Brunelleschi. Dix pour cent seulement des 186 ingénieurs espagnols du XVIe siècle dont les biographies furent étudiées par Garcia Tapia avaient une éducation scientifique[32].

On trouve la même rhétorique chez Agostino Ramelli (1531-après 1608) pour préciser dans son livre *Les diverses machines artificieuses* qui parut en 1588[33]. A la différence du mathématicien Pacioli, Ramelli était lui-même un ingénieur célèbre quand il atteignit quarante ans. Dans sa jeunesse, il avait par-ticipé aux expéditions militaires du marquis de Marignano, Gian Giacomo Medici, et avait étudié intensivement les mathématiques. En écrivant la préface de son oeuvre, il pouvait s'appuyer sur les mathématiciens de l'école d'Urbino, à savoir sur le " restaurateur des mathématiques " Federico Commandino et son élève, l'inspecteur général des forteresses toscanes, l'humaniste, le mathé-maticien et physicien Guidobaldo dal Monte[34]. S'il parle de la divinité des sciences mathématiques, du divin Archimède, s'il maintient que le seul et sûr fondement de tous les arts libéraux et mécaniques repose sur la vraie compré-hension des mathématiques, il nous rappelle toutefois de même les divinités mathématiques de Regiomontanus et de Pacioli.

Cependant, il ne faut pas s'illusionner sur la mécanique de Ramelli au regard des remarques enthousiastes. Sa mécanique n'analyse pas non plus la manière dont les machines fonctionnent en s'appuyant sur une base mathéma-tique. Lui-même ne les réalise pas non plus en prenant en considération les frictions et les dimensions appropriées au sens du génie moderne. Considérons à cette fin sa fantastique catapulte dont les dimensions sont gigantesques. (Illustration 20)

30. F. di Giorgio Martini, *Trattati di architettura ingegneria e arte militare*, *op. cit.* (voir note 2), vol. 2, 493.

31. H. Bredekamp, " Michelangelos Modellkritik ", B. Evers (éd.), *Architekturmodelle der Renaissance, Die Harmonie des Bauens von Alberti bis Michelangelo*, *op. cit.* (voir note 9), 116-123, voir 118 s.

32. G. Tapia, *Ingenieria y arquitectura en el Renacimento español*, Valladolid, 1990, 46-55, 63.

33. A. Ramelli, *Le diverse et artificiose machine*, Paris, 1588 (rééd. Milan, 1991).

34. A. Carugo, " Nota sulle fonti letterarie di Agostino Ramelli ", dans Ramelli, *loc. cit.* (voir note 33, Milan, 1991), XXXV-XLI.

ILLUSTRATION 20.

A. Ramelli, Une catapulte : A. Ramelli, *op. cit.* (voir note 33), figure 193.

Les trois soliveaux K, L, M sont composés en leur partie postérieure en forme de grandes cuillers ou auges, qui sont tirées vers le bas au moyen d'une manivelle et de cordes. De cette façon, le tréteau N, O qui pèse dessus est pressé vers le haut. Il accumule une force rotatrice au moyen de deux cordages tordus. Il est lié par une corde à la fin du soliveau P qui tient trois projectiles. Si on lâche le câble tendeur, les trois cuillers et le soliveau à trois bras lanceront — au moins théoriquement — leurs pierres et boules à droite.

L'exemple nous permet de préciser ce que ces auteurs s'imaginaient en soulignant l'importance des mathématiques comprises comme science des quantités : avant tout l'aspect quantitatif, non pas l'aspect fonctionnel des constructions, des machines. La géométrie décrit les figures et apprend à mesurer les hauteurs, les angles, les poids etc. De fait, Fontana, Taccola, Francesco dessinèrent beaucoup d'instruments et de méthodes de mesure[35].

Théorie et pratique

Ce que nous n'avons retrouvé que d'une manière indirecte chez les mathématiciens et ingénieurs mentionnés jusqu'ici, à savoir une technique théoriquement fondée, se trouve chez d'autres auteurs. L'ingénieur Léonard n'a pas

35. P.O. Long, " Power, Patronage, and the Authorship of Ars, From Mechanical Know-how to Mechanical Knowledge in the Last Scribal Age ", *Isis*, 88 (1997), 1-41, voir 11 ; P. Galluzzi, *op. cit.* (voir note 5), 122, 130-133.

seulement mis en évidence très souvent l'importance des mathématiques pour le génie. Il se proposait en fait d'élaborer une oeuvre Éléments des machines, dont les esquisses forment la partie essentielle du codex soi-disant Madrid I (1497-1499)[36]. Les frictions y jouent un rôle important et à cet égard, il était en fait de loin en avance sur son temps. Mais même lui, il n'a pas fait le pas crucial du dessin des machines à leur réalisation : ses esquisses demeurèrent justement ce qu'elles étaient, à savoir des esquisses.

Le mathématicien Tartaglia, le père de la balistique, et contemporain plus jeune de Léonard, affirma pouvoir donner une fondation théorique de ses règles balistiques. Son Invention découverte récemment s'appelant Nouvelle science s'adressa à des mathématiciens, des bombardiers et à d'autres. Sans avoir aucune pratique d'artilleur, il prétendit savoir pour des raisons évidentes que deux angles différents de lancement entraînent la même portée. Il ajouta à bon droit que c'était un problème auquel personne n'avait pensé jusque-là[37]. La question concerne le problème inverse dans un milieu résistant : étant donné une portée qui est plus petite que la maximale, il y a exactement deux· angles de lancement. La preuve mathématique de cette affirmation nécessite la solution d'une équation différentielle linéaire et était par conséquent de loin hors d'atteinte de Tartaglia.

Peut-être attend-on une corrélation plus étroite entre les deux disciplines lorsqu'il s'agissait d'un ingénieur qui était en même temps l'un des mathématiciens les plus importants du XVIe siècle, à savoir de Raffael Bombelli. Loin de là ! En tant qu'ingénieur, il était en service chez son patron Alessandro Ruffini et défricha le marais du val toscan Chiana. Dans la préface de son livre sur l'algèbre, écrit pendant une interruption des travaux, il rapporte fièrement que tous les gens apprécient ce travail de génie comme glorieux et immortel, comme une *opera … gloriosa, ed immortale*[38]. L'ingénieur se flattait d'une façon qui rappelle évidemment le vers fameux de Horace[39] :

Exegi monumentum aere perennius.

J'ai achevé un monument qui est plus durable que le bronze.

En fait, Bombelli était en même temps un humaniste qui traduisit avec Antonio Maria Pazzi cinq des sept livres de Diophante du grec en latin et réorganisa son algèbre d'après Diophante[40].

36. H. Maschat, *Leonardo da Vinci und die Technik der Renaissance*, Munich, 1989.

37. C.W. Groetsch, " Tartaglia's Inverse Problem in a Resistive Medium ", *American Mathematical Monthly*, 103 (1996), 546-551.

38. R. Bombelli, *L'algebra opera con la quale ciascuno da se potrà venire in perfetta cognitione della teorica dell'Arimetica*, Bologne, 1579, Préface dédiée à A. Ruffini, 3.

39. Horace, *Odes III*, 30, vers 1.

40. P.L. Rose, *The Italian Renaissance of mathematics, Studies on humanists and mathematicians from Petrarch to Galileo*, Genève, 1975, 145 s.

Il n'y avait aucun lien entre ses activités en tant qu'ingénieur et ses activités en tant que mathématicien. Ce sont seulement le plus jeune contemporain Simon Stevin (1548-1620), qui reçut cinq brevets pour quinze inventions, et plus tard Johannes Faulhaber (1580-1625), tous les deux des mathématiciens et des ingénieurs, qui ont lié systématiquement la théorie mathématique à la pratique de l'ingénieur, qui ont développé des cours d'instruction pour des ingénieurs instruits en mathématiques[41].

LES DESCENDANTS D'ARCHIMÈDE

Dans sa vaste conception d'architecte, Vitruve avait réuni l'architecte et le machiniste en une seule personne, de sorte qu'encore Fausto Veranzio commença sa déclaration des machines en disant que " plusieurs ont tenu que l'art des machines estoyt le principal dans l'architecture "[42]. En effet, la Renaissance vitruvienne ouverte par Leon Battista Alberti jouait un rôle essentiel aussi pour le génie.

Les auteurs des traités correspondants citaient Vitruve volontiers et souvent, avant tout Francesco di Giorgio Martini dans ses *Trattati di architettura ingegneria e arte militare*. Toutefois Archimède était le vrai modèle de la figure d'identification de l'ingénieur de la Renaissance. Ses écrits avaient été traduits plusieurs fois en latin jusqu'au XVIᵉ siècle[43]. Tous ses écrits concernaient les mathématiques pures, la statique ou l'hydrostatique, mais pas la mécanique appliquée, pas la mécanologie, pas l'art militaire. Archimède n'en a laissé aucune note comme Plutarque nous fait savoir dans sa biographie du général romain Marcellus[44]. Cependant, il avait fondé sa gloire légendaire comme ingénieur militaire par ses machines défensives. Il était le *machinator praestantissimus*, le constructeur de machines le plus célèbre, comme le grand chercheur des écrits archimédiens, Francesco Maurolico, l'appelait au XVIᵉ siècle[45], " un des mécaniciens les plus importants ", comme l'admirateur d'Archimède, Guidobaldo dal Monte, dit en 1577.

41. H. Schimank, *Der Ingenieur, Entwicklungsweg eines Berufes bis Ende des 19. Jahrhunderts*, Cologne, 1961, 24.

42. F. Veranzio, *Machinae novae*, Venise, 1615/16 (rééd. Munich, 1965).

43. M. Clagett, " The medieval Archimedes in the Renaissance, 1450-1565 ", *Archimedes in the Middle Ages*, vol. 3, Madison, 1964, Philadelphia, 1976-1980, 4 vols ; I. Schneider, *Archimedes*, Darmstadt, 1979, 164-168.

44. Plutarch, " Grosse Griechen und Römer ", *Marcellus*, vol. 3, eingeleitet und übersetzt von Konrat Ziegler, Zurich, Stuttgart, 1955, 302-344, voir 321.

45. M. Clagett, *op. cit.* (voir note 43), vol. 3, 813 ; G. Ubaldo Marquis del Monte, *Mechanicorum liber*, Pesaro, 1577, With the commentaries of F. Pigafetta from his Italian translation *Le Mechaniche Venice*, 1581, Abridged translation from the Italian with notes by Stillman Drake, dans *Mechanics in Sixteenth-Century Italy, Selections from Tartaglia, Benedetti, Guido Ubaldo, & Galileo*, translated and annotated by St. Drake and I.E. Drabkin, Madison, Milwaukee, London, 1969, 239-328, voir 256.

Archimède était le synonyme pour " fertilité en inventions ", pour la faculté de trouver de nouvelles solutions. Faut-il s'étonner que les auteurs innombrables de manuscrits illustrés et de livres des machines de la Renaissance se soient référés à Archimède comme aïeul, quand ils soulignaient constamment qu'ils présentaient des machines nouvelles, utiles, fertiles en inventions ? Être ingénieur voulait dire être fertile en inventions, disposer d'ingenium, de la sagacité. La fameuse sentence archimédienne *Eurêka, eurêka*, j'ai trouvé, j'ai trouvé, transmise par Vitruve[46], caractérise exactement cet aspect. En conséquence, Pacioli célébra les *perspicacissimi architecti e ingegnieri e di cose nove assidui inventori*, dans son écrit *Divina proportione*, les architectes et ingénieurs et inventeurs assidus de choses nouvelles, qu'on peut rencontrer à la cour du duc milanais[47].

C'est la raison pour laquelle la plus grande louange, la plus grande gloire d'un ingénieur consistait à égaler Archimède. Taccola se flattait d'être nommé Archimède de Sienne[48]. Ambrose Bachot appelait Agostino Ramelli l'Archimède de son âge[49].

Le comte Guillaume IV de Hesse-Cassel appelait son horloger et fabricant d'instruments Jost Bürgi (1552-1631) un deuxième Archimède. En fait, il ne faut pas oublier que le poète romain Claudianus avait glorifié le planétaire d'Archimède[50], que Brunelleschi aussi est un autre horloger. On pourrait continuer cette série.

Par conséquent, tous les ingénieurs éminents de la Renaissance citaient Archimède, ce divin Archimède selon Agostino Ramelli[51] : Giovanni Fontana, Francesco di Giorgio Martini et Léonard, Jérôme Cardan, Pedro Juan de Lastanosa (vers 1564/75)[52], Bunaiuto Lorini (1597), Heinrich Zeising (1612/14), Jacques Besson (1569) qui s'appela lui-même " docte mathématicien " sur la page de titre de son *Théâtre des instruments mathématiques et mécaniques*.

---

46. Vitruve, *op. cit.* (voir note 3), IX préface § 10.

47. L. Pacioli, *op. cit.* (voir note 28), 33, 181.

48. M. Taccola, *op. cit.* (voir note 16), 53.

49. G. Scaglia, " Introduzione - Introduction ", dans A. Ramelli, *op. cit.* (voir note 33, Milan 1991), IX-XXXIII, voir XII, XXIV.

50. Claudianus, " In sphaeram Archimedis ", *Carmina minora*, 51.

51. A. Ramelli, *op. cit.* (voir note 33), préface p. IIIV ; E. Battisti et G. Saccaro Battisti (éds), *op. cit.* (voir note 15), 61 s. ; F. di Giorgio Martini, *op. cit.* (voir note 2), vol. 1, 6 ; P. Galluzzi, *op. cit.* (voir note 5), 63 ; G. Cardan, *De subtilitate*, Nuremberg, 1550 (rééd. dans *Opera omnia*, vol. 3, in C. Spon (ed.), Lyon, 1663), 352-672, voir 607.

52. Pseudo-Juanelo Turriano, *The twenty-one books of devices and of machines*, vol. 1, General introduction by G.A. García-Diego, Madrid, 1984 (rééd. Madrid, 1997), 2 vols, 108 ; B. Lorini, *Delle Fortificationi*, Venise, 1597, 172 ; H. Zeising, *Theatrum machinarum*, partie 1, Leipzig, 1612-1614, 27, 6 parties ; M. Popplow, *Wandlungen des Technikverständnisses im Übergang vom Mittelalter zur Neuzeit* (sous presse), chapitre " Mathematik und Mechanik als theoretisches Fundament des Entwerfens von Maschinen ", notes 65, 67 ; J. Besson, *Théâtre des instruments mathématiques et mécaniques*, Lyon, 1596.

L'éditeur François Beroald de la nouvelle édition posthume qui parut en 1596 souligna dans sa préface pour le lecteur : " Le tout [est] inventé de tel esprit que les plus doctes Mathématiciens certifioyent n'y avoir eu iamais inventions mathématiques plus proffitables ". Cette remarque inclut les machines archimédiennes.

La figure 54 représente le bateau et les trois vis sans fin avec lesquels Archimède tira un vaisseau de merveilleuse grandeur de terre en mer.

(Illustration 21)

ILLUSTRATION 21.

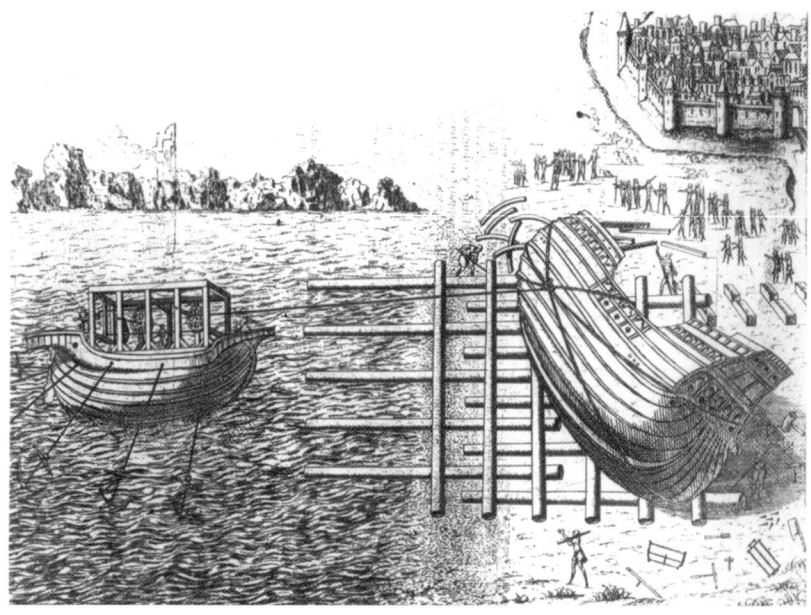

J. Besson, Un bateau est tiré en mer : J. Besson,
op. cit. (voir note 52), figure 54.

Parmi nombreux auteurs qui ont choisi Archimède comme figure symbolique sur le frontispice de leurs livres de construction ou de machines, on trouve les cinq suivants :

1. Samuel Marolois et ses *Opera mathematica ou oeuvres mathématiques traictans de géométrie, perspective, architecture, et fortification*, La Haye, 1614[53].

___

53. S. Marolois, *Opera mathematica ou oeuvres mathématiques traictans de géométrie, perspective, architecture, et fortification*, La Haye, 1614 ; H. Neumann, " Architectura militaris ", dans U. Schütte (éd.), *Architekt und Ingenieur, Baumeister in Krieg und Frieden*, Wolfenbüttel, 1984, 281-404, voir 363.

Euclide, Vitellion, Vitruve et Archimède représentent la géodésie, la navigation et l'astronomie, l'architecture, la science militaire. (Illustration 22)

ILLUSTRATION 22.

S. Marolois, *Opera mathematica,* etc.
(La Haye, 1614), page de titre.

2. Salomon de Caus et son livre *Sur les mouvements violents. Description
de quelques machines utiles aussi bien que divertissantes et de différents précis
de quelques cavernes ou grottes et fontaines divertissantes*, Francfort-sur-le-
Main, 1615.

Archimède et Héron d'Alexandrie représentent ensemble le génie, Archimède l'utilité par un ressort, une moufle, une balance, Héron le divertissement
par un tuyau courbé pour une fontaine. (Illustration 23)

ILLUSTRATION 23.

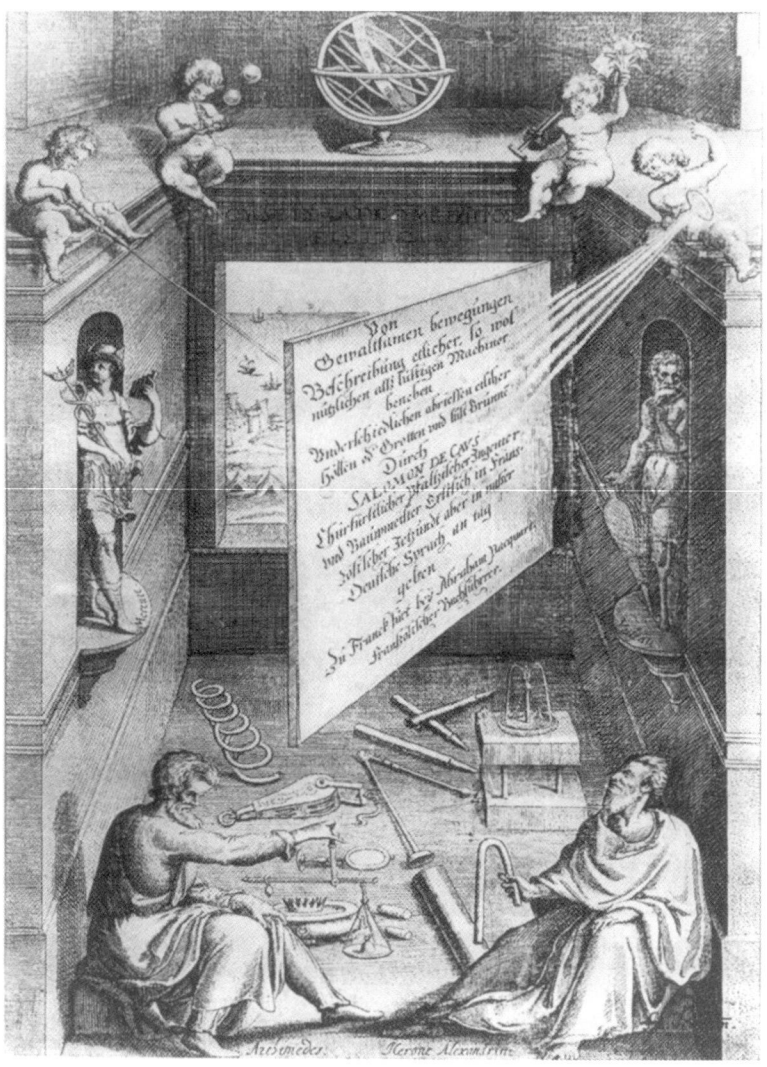

S. de Caus, *Von Gewaltsamen bewegungen,* etc.
(Francfort-sur-le-Main, 1615).

3. Jacopo Strada et son *Précis habile de quelques moulins à eau, à vent, à cheval et à main ainsi que de pompes belles et utiles et d'autres machines pour élever l'eau et de fontaines divertissantes et d'usines hydrauliques qu'on n'a jamais vues auparavant,* Francfort-sur-le-Main, 1617. (Illustration 24)

ILLUSTRATION 24.

J. Strada, *Kunstliche Abriß*, etc.
(Francfort-sur-le-Main, 1617), page de titre.

Archimède et Vitruve représentent le génie et l'architecture[54]. La même chose s'applique à :

54. J. Strada, *Kunstliche Abriß/allerhand Wasser- Wind- Roß- und Handt Mühlen/beneben schönen und nützlichen Pompen/auch andern Machinen/damit das Wasser in Höhe zuerheben/ auch lustige Brunnen und Wasserwerck/dergleichen vor diesem nie gesehen worden. Nicht allein den Liebhabern zur Ubung und Nachrichtung/sondern auch dem gantzen gemeinen Vaterland/zu Dienst und Wolgefallen/so wol in Kriegs- als Friedens zeiten zugebrauchen*, Francfort-sur-le-Main, 1617.

4. Giovanni Branca et ses *Machines, Un nouveau volume, fait avec beau-coup d'art pour réaliser des effets merveilleux spirituels aussi bien que par une opération animale*, Rome, 1629. (Illustration 25)

ILLUSTRATION 25.

G. Branca, *Le machine,* etc.
(Rome 1629), page de titre.

5. Georg Andreas Böckler et son *Nouveau théâtre des machines, Théâtre des arts mécaniques de moulins et d'usines hydrauliques*, Nuremberg, 1661[55].

Archimède et un mécanicien lèvent le rideau d'un théâtre et rendent possible ainsi un regard sur des moulins et des usines hydrauliques. Les titres " étude " (*studium*) et " travail " (*labor*) caractérisent les deux comme des représentants du génie théorique et pratique. (Illustration 26)

55. S. de Caus, G. Branca, G.A. Böckler : voir note 4 ; A. Stöcklein, *Leitbilder der Technik, Biblische Tradition und technischer Fortschritt*, Munich, 1969, 177.

ILLUSTRATION 26.

G. A. Böckler, *Theatrum machinarum novum*, etc.
(Nuremberg 1661), page de titre.

FANTAISIE ET UTOPIE

Les pages de titre indiquent en même temps quatre aspects importants de cette littérature d'ingénieurs : divertissant, utile, nouveau, merveilleux. Ces aspects méritent d'être considérés un peu plus en détail.

Divertissant

La technique divertissante appartenait au ressort de l'ingénieur depuis l'antiquité : les instruments de musique, les automates, la plantation, la sauvegarde et la décoration de jardins d'agrément, les jardins demandaient souvent de grandes dépenses de technique[56]. Prenons par exemple les parcs magnifi-

56. A. Stöcklein, *op. cit.* (voir note 55), 53-56.

ques d'Aranjuez, de la résidence de printemps des rois espagnols qui se trouve près de Madrid ou les jardins autour du château monastique San Lorenzo del Escorial, de la résidence d'été et d'automne de ces rois qui fut construit par Philippe II entre 1563 et 1584. Francesco dessina des jardins[57]. La même chose s'applique au Manuel du moyen-âge comme nous avons vu plus haut. Chez Branca on trouve un organiste[58]. Vers 1420, Giovanni Fontana dessina des orgues, des jongleurs, des masques théâtraux et des labyrinthes. Il mentionna même qu'il avait dessiné un livre sur cinq formes de labyrinthes suivant sa propre fantaisie et renvoya à Dédale[59]. Le nom de cet inventeur mythique du second millénaire avant l'ère chrétienne ne veut rien dire d'autre que " l'artiste " qui a créé d'après un on-dit aussi des statues se mouvant automatiquement. En conséquence, en 1446, l'humaniste Carlo Marsuppini fit l'éloge de Brunelleschi sur sa pierre tombale : ses nombreuses machines peuvent témoigner des capacités de Filippo au regard de son esprit divin dans l'art de Dédale[60]. On se rappelle l'interprétation des deux cavaliers sur le tableau de Piero di Cosimo. En tant que constructeur de labyrinthes, Dédale était en même temps l'aïeul des architectes de sorte que Ambrose Bachot nomma Ramelli un vrai Dédale en tant qu'architecte[61].

Utile et nouveau

Nouveauté, utilité, applicabilité, novità, utilità, idonietà étaient les trois propriétés que les autorités italiennes examinaient avant qu'elles accordassent les privilèges d'invention. En conséquence, les auteurs les indiquèrent dans les titres de leurs livres et descriptions de machines. Venise régla l'adjudication de brevets à des inventeurs d'innovations techniques en 1477 par la *Parte Veneziana sulle Invenzioni*. On examinait principalement l'utilité et l'applicabilité. " Nouveau " voulait seulement dire que l'invention était inconnue à la région vénitienne[62]. A vrai dire, il ne s'agissait souvent que de modifications peu considérables de dispositifs techniques disponibles.

Mais c'était justement ce en quoi le maître se manifestait. Taccola dit expressément[63] : " Tout maître dans son art recherchera l'avantage de son ouvrage. C'est alors seulement qu'il pourra être appelé un maître es arts mécaniques ". (Illustration 27)

57. F. di Giorgio Martini, *op. cit.* (voir note 2), vol. 1, tableau 45.

58. G. Branca, *op. cit.* (voir note 4), figure 21.

59. E. Battisti et G. Saccaro Battisti, *op. cit.* (voir note 15), 61.

60. E. Battisti, *Filippo Brunelleschi*, Milan, 1976, 16 ; M. Kemp, *op. cit.* (voir note 2), 394 ; P. Galluzzi, *op. cit.* (voir note 5), 24.

61. G. Scaglia, *op. cit.* (voir note 49), XII, XXIV.

62. H. Schippel, " Die Anfänge des Erfinderschutzes in Venedig ", dans U. Lindgren (éd.), *loc. cit.* (voir note 13), 539-550.

63. E. Knobloch (éd.), *L'art de la guerre, Machines et stratagèmes de Taccola, ingénieur de la Renaissance*, Paris, 1992, 131, 202.

ILLUSTRATION 27.

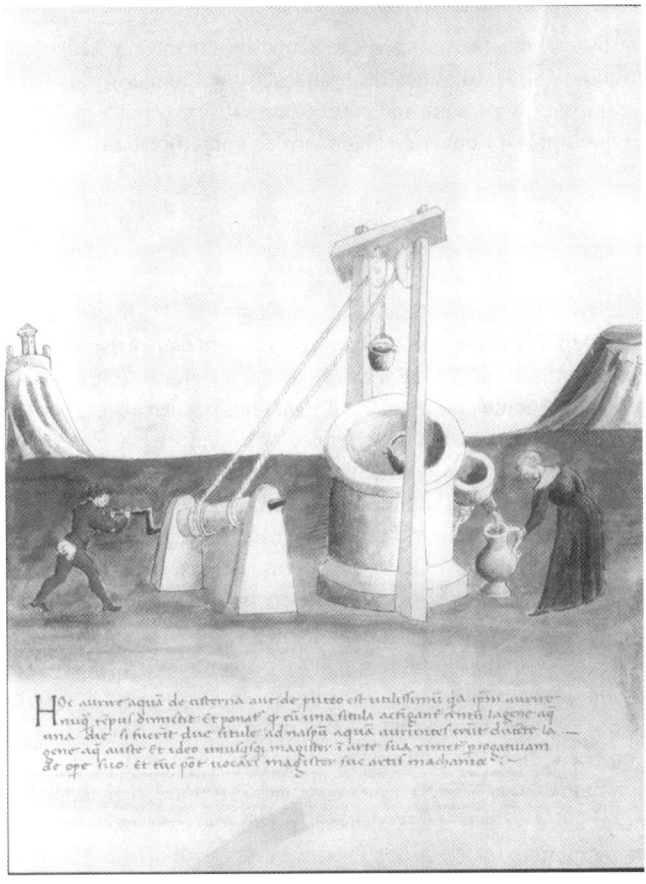

Mariano Taccola, Un puits : *L'art de la guerre, op. cit.* (voir ill. 8), 131.

Son dessin montre comment on peut puiser la double quantité d'eau si l'on met au treuil deux seaux au lieu d'un seul. Étant donné le grand nombre de machines utilisées dans l'architecture depuis l'Antiquité, la gloire devait être d'autant plus grande pour ceux qui inventèrent de nouvelles machines des centaines d'années plus tard, comme Fausto Veranzio constata au début du XVIIᵉ siècle[64]. Il va de soi que l'ouvrage Sur les inventeurs des choses par Polidoro Vergilio qui parut en 1499 pour la première fois était très populaire à une époque qui estimait tellement la fertilité en inventions techniques et la créativité artistique[65]. Il y en a eu 55 éditions ou traductions.

64. F. Veranzio, *op. cit.* (voir note 42), *Déclaration des machines.*
65. P. Vergilio, *De rerum inventoribus*, Bâle, 1540.

La notion mentionnée de " nouveauté " facilitait la possibilité, il est vrai, de s'approprier des inventions étrangères.

Les avertissements mentionnés de Brunelleschi envers Taccola en témoignent, de même que les plaintes de Francesco mentionnées plus haut, dont les dessins furent copiés presque inchangés par Vittorio Zonca et Jacopo Strada. Fausto Veranzio prévit l'envie de beaucoup de contemplateurs de ses *Nouvelles Machines*.

Merveilleux

Dès le début, il y eut une relation étroite entre la technique et le monde du miraculeux. Déjà les *Problèmes mécaniques* attribués à Aristote avaient souligné l'étonnement causé par la manière d'opérer des machines. Son édition latine par le constructeur humaniste de bateaux, Vettor Fausto, était disponible depuis 1517[66] et était souvent mentionnée entre autres par Ramelli et Branca. Cardan parlait des inventions miraculeuses (*mira inventa*) d'Archimède, de cet auteur inimitable[67]. Pour l'admirateur d'Archimède Guidobaldo dal Monte, une machine avait le caractère d'une merveille[68]. Ramelli et Branca soulignèrent les effets merveilleux de leurs machines, Branca déjà dans le titre de son livre. En fait, les machines suggèrent une image de l'ingénieur très proche de celle du magicien[69]. Cette tradition ne cessa pas au XVIIe ou XVIIIe siècle. Originairement Caspar Schott voulait appeler sa " Magie universelle " *Thaumaturgus physico-mathematicus*, " qui accomplit des miracles physico-mathématiques "[70]. Sa " Magie accomplissant des miracles " décrit des machines compliquées, en particulier des automates comme le pigeon d'Archytas ou l'aigle de Regiomontanus. Sa " Technique curieuse " répond des *mirabilia artis*[71], " des admirables de l'art ". Ses dessins posthumes de machines artificielles (*Machinae artificiales*) varient des machines de la Renaissance[72]. Il s'agit des moulins, des pompes, des grues, dont les roues dentées, les transmissions sont multipliées d'une manière baroque : (Illustrations 28-30).

66. F. Klemm, " Die Rolle der Technik in der italienischen Renaissance ", *Technikgeschichte*, 32 (1965), 221-243, voir 238.

67. G. Cardan, *op. cit.* (voir note 51), 607.

68. F. Klemm, *op. cit.* (voir note 66), 240.

69. P. Galluzzi, *op. cit.* (voir note 5), 26.

70. C. Schott, *Magia universalis naturae et artis opus quadripartitum*, Würzburg 1657-1659, 4 vols.

71. L. Boehm, " Historische Staats-Metaphorik aus Natur und Technik - rhetorische Stilistik oder Weltanschauung ? Wissenschaftsgeschichtliche Überlegungen zum Interpretationswandel ", *Schriften der Georg-Agricola-Gesellschaft*, 18 (1992), 13-40, voir 34.

72. Machinae Artificiales R.P. Gasparis Schotti Soc.is Jesu Mathematici Herbipolensis : Handschriftenabteilung der Bibliothek der Universität zu Würzburg, signature Delin. 5 (vers 1650) ; la reproduction des dessins est approuvée par la bibliothèque de l'Université de Würzburg.

ILLUSTRATION 28.

C. Schott, Un moulin double commandé par énergie hydraulique :
Machinae artificiales (voir note 72).

ILLUSTRATION 29.

C. Schott, Deux pompes avec une commande par énergie hydraulique :
Machinae artificiales (voir note 72).

ILLUSTRATION 30.

C. Schott, Un arbre manivelle pour commander une grue :
Machinae artificiales (voir note 72).

La confiance dans les possibilités presque illimitées de la technique se reflétait dans les nombreux projets fantastiques des ingénieurs de la Renaissance[73]. " Fantaisie " était un mot-clé de la théorie de l'art, de l'architecture et de la poésie à la Renaissance liée étroitement à la notion de créativité[74].

Non seulement Fontana, mais aussi Filarete soulignèrent son rôle. Pour lui, la fantaisie était l'impulsion initiale d'une oeuvre. C'était l'utopie qui lui correspondait dans la théorie de l'architecture[75]. Par conséquent, Léon Battista Alberti décrivit sa société utopique, Filarete sa cité utopique Sforzinda dans leurs traités humanistes de l'architecture. Voici une telle utopie d'un peintre inconnu[76].

73. P. Galluzzi, *Les ingénieurs de la Renaissance de Brunelleschi à Léonard de Vinci, op. cit.* (voir note 5), 47 ; J. Gimpel, *La révolution industrielle du Moyen Age, op. cit.* (voir note 27), 222.

74. M. Kemp, " From " Mimesis " to " Fantasia " : the Quattrocento vocabulary of creation, inspiration and genius in the visual arts ", *op. cit.* (voir note 2), 361-376.

75. *Idem,* 370.

76. M. Herrmann, " Die Utopie als Modell ", Zu den Idealstadt-Bildern in Urbino, Baltimore und Berlin ", B. Evers (éd.), *Architekturmodelle der Renaissance, Die Harmonie des Bauens von Alberti bis Michelangelo, op. cit.* (voir note 9), 56-73, voir 66 s., 199 s.

ILLUSTRATION 31.

Peintre anonyme, Une cité idéale (vers 1480) :
M. Herrmann, *op. cit.* (voir note 76), 200.

D'après la signification du mot grec fantaisie était l'imagination. Son usage n'était pas toujours sans contradiction au XVᵉ siècle. Mais il restait associé à la procédure rationnelle et intellectuelle de l'invention. Giorgio Vasari louait la fertilité en inventions de Piero di Cosimo et de Léonard de Vinci dans ses Vies d'artistes (1550)[77]. Et en fait, ce fut particulièrement Léonard, et aussi la plupart des autres ingénieurs qui imaginèrent des techniques qui n'étaient — peut-être pas encore — du domaine du possible. Considérons trois exemples :

Un scaphandrier avec l'équipement nécessaire (un masque subaquatique avec tube respiratoire etc.) pour pouvoir aller au fond des eaux comme déjà Kyeser le dessina en 1405 dans son Bellifortis (" Le fort en combat "), plus tard l'Anonyme de la Guerre hussite, Taccola ou Léonard de Vinci[78].

Les mouvements perpétuels qui apparaissent chez Taccola, Francesco, Léonard et encore chez Böckler et Strada[79] : Son mécanisme même permet d'aiguiser des marteaux.

77. Sh. Fermor, *Piero di Cosimo, Fiction, Invention and Fantasia*, London, 1993, 26.

78. C. Kyeser, *Bellifortis, op. cit.* (voir note 12), vol. 1, 40 et vol. 2, f. 62 r ; Bert S. Hall (éd.), *The technological illustrations of the so-called " Anonymous of the Hussite war "*, codex Latinus Monacensis 197, Part 1, Wiesbaden, 1979 ; P. Galluzzi, *op. cit.* (voir note 5), 124-126.

79. F. Klemm, " Physik und Technik in Leonardo da Vincis Madrider Manuskripten ", dans F. Klemm (éd.), *Zur Kulturgeschichte der Technik, Aufsätze und Vorträge 1954-1978* (Kulturgeschichte der Naturwissenschaften und der Technik, 1), Munich, 1979, 109-138, voir 113 ; E. Rupp (éd.), *Mechanismen des 18. Jahrhunderts*, Heidelberg, 1970, 22-27 ; A. Stöcklein, *op. cit.* (voir note 55), 183.

Un char à quatre roues qui était conçu comme une vraie automobile au strict sens du mot de sorte qu'il se mouvait de lui-même. (Illustration 32)

ILLUSTRATION 32.

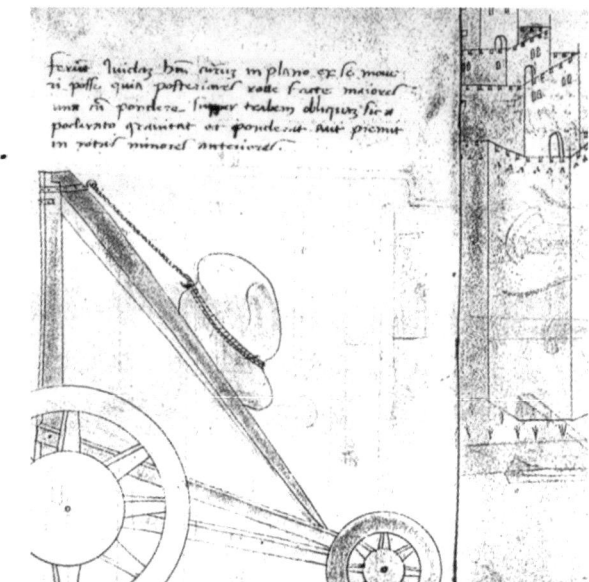

G. Fontana, Une vraie automobile :
E. Battisti et G. Saccaro Battisti, *op. cit.* (voir note 15), 78.

Giovanni Fontana le dessina[80]. Évidemment sa description est quand même sceptique : " Quelques-uns rapportent que ce char peut se mouvoir dans la plaine parce que les roues arrières sont plus grandes ". Le poids se trouvant sur une planche inclinée pèse ou presse sur les roues plus petites de devant.

CONCLUSION : LE TRIOMPHE DE L'ESPRIT

L'ingénieur est l'homme qui accomplit des miracles. Nous l'avons entendu. Encore en 1725, l'ingénieur Jakob Leupold de Leipzig expliqua dans son Théâtre des machines consistant en huit volumes : " Aujourd'hui, nos ingénieurs sont ce qu'étaient ces mécaniciens il y a longtemps. Il ne leur incombe pas seulement la tâche de tracer et de construire une forteresse, mais aussi de

80. E. Battisti et G. Saccaro Battisti, *op. cit.* (voir note 15), 78.

décrire différentes machines selon les fondements mécaniques…, d'inventer des machines, de faciliter le travail et de rendre quand même possible ce qui semble être assez souvent impossible "[81]. Sa fertilité en inventions et sa sagacité le rend à même de le faire, symbolisent le triomphe de l'esprit. En conséquence, les ingénieurs glorifiaient constamment l'ingenium, ce qui voulait dire la sagacité aussi bien que l'invention spirituelle.

Taccola appréciait de telles sentences : "A partir d'une petite invention ingénieuse, on peut soulever des objets assez lourds ". Cette phrase commente le dessin d'une moufle avec un treuil (Illustration 33). La phrase " Une invention astucieuse est plus efficace que la force des boeufs " commente le dessin d'un banc à quatre pieds muni d'une vis[82].

ILLUSTRATION 33.

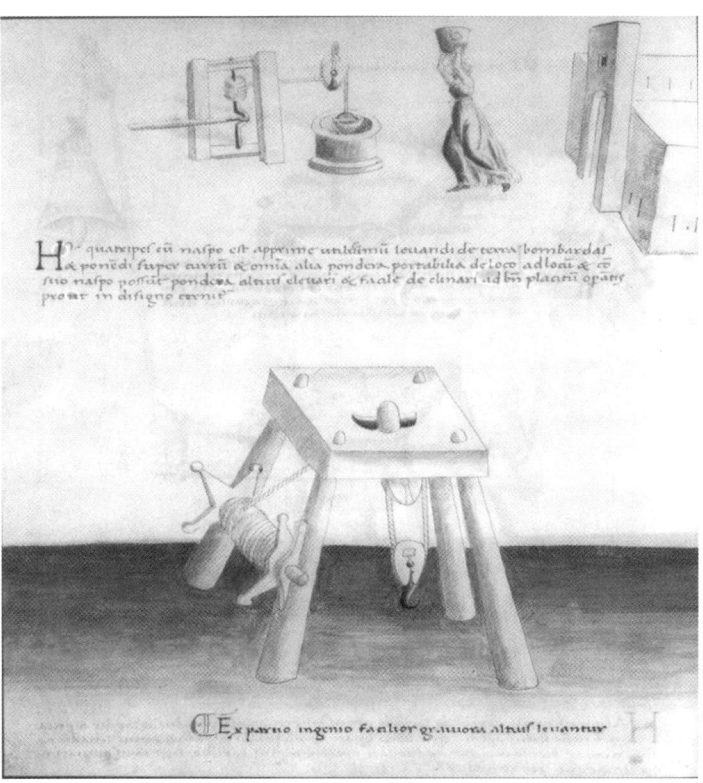

Mariano Taccola, Une sentence :
l'art de la guerre, *op. cit.* (voir ill. 8), 110.

---

81. J. Leupold, *Theatrum machinarum oder : Schau-Platz der Heb-Zeuge*, Leipzig, 1725 (rééd. Hanovre, 1982), Dédication p. 3 ; E. Rupp, *op. cit.* (voir note 79), 19.

82. *L'art de la guerre, op. cit.* (voir illustration 8), 110 s.

L'aphorisme *Aurum probatur igni*, et *ingenium mathematicis*, " L'or est éprouvé par le feu et la sagacité par les mathématiques " orna le frontispice de la *Nouvelle Science* de Tartaglia en 1537[83]. On le retrouve dix ans plus tard chez Guillaume Ryff[84]. Son pentamètre *Vivitur ingenio, caetera mortis erunt*, " On vit par l'esprit, le reste est perdu " est utilisé de nouveau en 1584 par Jean Errard de Bar-Le-Duc dans son *Premier livre des instruments mathématiques et mécaniques*[85]. (Illustration 34)

ILLUSTRATION 34.

G. Ryff, *Der furnembsten notwendigsten der gantzen Architectur angehörigen Mathematischen und mechaniscen Künst eygentlicher bericht,* etc. (Nuremberg, 1547), deuxième page de titre avec une sentence.

---

83. *Mechanics and Sixteenth-Century Italy, op. cit.* (voir note 45), 18.

84. E. Knobloch, " Praktische Geometrie ", M. Folkerts, E. Knobloch, K. Reich (éds), *Maß, Zahl und Gewicht, Mathematik als Schlüssel zu Weltverständnis und Weltbeherrschung*, Wolfenbüttel, 1989, 123-154, voir 148.

85. A. Stöcklein, *op. cit.* (voir note 55), 137 note 125.

Le déplacement et le transport des charges les plus lourdes comme de parties de bâtiments ou d'obélisques comptaient au nombre des travaux les plus spectaculaires des ingénieurs de la Renaissance. Francesco dessina plusieurs machines à de tels déplacements[86] qui avaient rendu célèbre en particulier l'architecte, ingénieur et fondeur en bronze Aristotele di Fioravante (né environ 1425). Le pape Paul II lui confia la tâche de déplacer l'obélisque énorme se trouvant sur la place Saint Pierre. Mais le pape mourut. Seulement Sixtus V fit réaliser ce travail par l'architecte et ingénieur papal Domenico Fontana.

Fontana en publia un rapport quatre ans plus tard, qui donne une idée claire de ce qui s'est passé[87]. L'obélisque était haut de 25, 35 mètres, avait un volume d'environ 125 m3 (mètres cubes), pesait 319 tonneaux métriques et devait être déplacé d'environ 257 mètres. La technique proposée par Fontana fut choisie parmi beaucoup d'autres dont il en décrivit sept sur cette illustration (Illustration 35).

ILLUSTRATION 35.

D. Fontana, *Del modo tenuto nel trasportare l'obelisco Vaticano*, etc.
(voir note 87), 8.

86. P. Galluzzi, *op. cit.* (voir note 5), 164-168.

87. Domenico Fontana, *Del modo tenuto nel trasportare l'obelisco Vaticano e delle fabriche fatte da nostro Signore Sisto V. libro primo*, Rome, 1590 (rééd. en 2 vols. par D. Conrad, Berlin 1987) ; F. Klemm, " Die Technik der italienischen Renaissance ", dans F. Klemm (éd.), *Zur Kulturgeschichte der Technik*, *loc. cit.* (voir note 79), 96-108, voir 106 s.

B   L'obélisque est transporté se tenant debout. Des soliveaux auxiliaires se trouvant à la pointe.

C   Il est tenu en équilibre sur une demi-roue.

D   Il n'est tenu que par des coins.

E   Il est mis sur le côté au moyen de vis et transporté en position inclinée.

F   Il est levé par un levier d'après le principe de la balance.

G   Il est incliné par une demi-roue dentée, une dent après l'autre.

H   Il est levé, incliné et tiré seulement par des vis.

Fontana proposa de lever l'obélisque, de le mettre par terre, de le tirer sur des rouleaux à la nouvelle place et de l'y relever. Comme l'illustration le montre, il disposait seulement de moyens très simples, mais cela en nombre suffisamment élevé. Un échafaudage gigantesque servait de dispositif de retenue. 40 moufles furent mues pour le mettre par terre. 907 ouvriers et 75 chevaux collaborèrent. 800 hommes et 140 chevaux le relevèrent. (Illustration 36)

ILLUSTRATION 36.

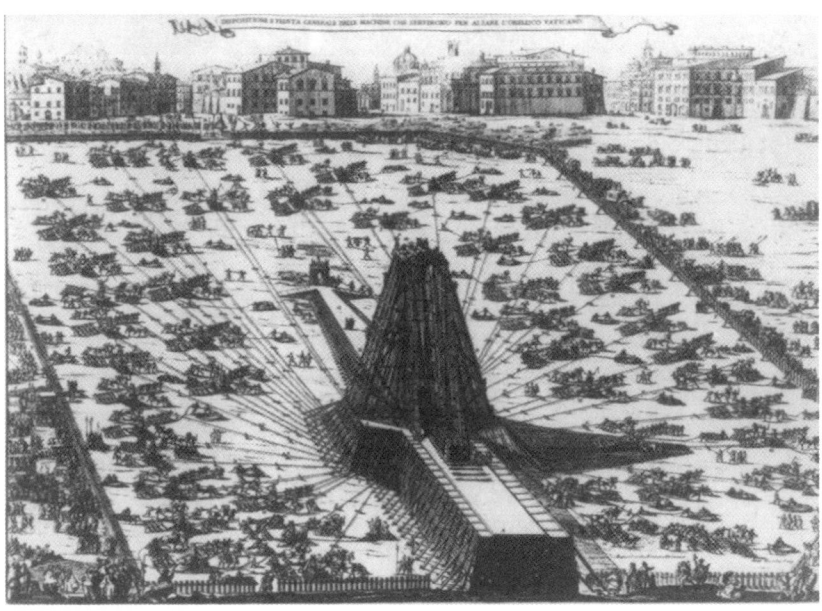

Carlo Fontana, Le relèvement de l'obélisque vatican :
Th. Hänseroth, " Transport und Versetzen schwerer Lasten von frühesten
Zeiten bis zum Ende des 16. Jahrhunderts "
dans D. Fontana, *Del modo tenuto nel trasportare l'obelisco vaticano*, etc.,
rééd. par D. Conrad, vol. " Traduction et commentaires ", 62-76, voir 73.

Un privilège papal lui garantit les pleins pouvoirs et les moyens adéquats. Il calcula les nombres nécessaires en calculant le volume de l'obélisque, le poids spécifique du matériau moyennant un cube à titre d'essai, ensuite le poids complet. Il savait par expérience la puissance d'un treuil avec de bons câbles et de bonnes moufles.

Une organisation astucieuse des travaux auxquels il ne manquait pas le moindre détail lui permit de réaliser avec succès sa tâche. Il remplit la même tâche à Rome plusieurs fois et déplaça trois autres obélisques. Quoiqu'il y ait une analogie évidente entre Brunelleschi et Fontana sur ce point, il y a aussi une différence cruciale : tandis que Brunelleschi s'appuya sur une innovation avec laquelle personne n'avait aucune expérience, dont le risque était en conséquence complètement inconnu et ne pouvait pas être mesuré ou calculé, Fontana s'appuya sur des moyens traditionnels, dont la coordination était, il est vrai, le plus grand problème, mais qui rendit le risque calculable. A cet égard, lui aussi fut un initiateur de la mécanique technique fondée sur les mathématiques.

Hydraulics in the late Renaissance 1550-1625.
Mathematicians' involvement in hydraulic engineering and the "mathematical architects"

Cesare S. Maffioli

Introduction

Although seventeenth-century hydraulics was a science in its own right, historians have usually looked at it through bifocal lenses which have split its image into two mutually irreconcilable parts. Some have shifted the focus forward to the mathematical science of fluids in the age of Euler, while others have almost exclusively directed attention backward to Leonardo's hydraulic activity in the early Renaissance. In this paper I would like to challenge this dichotomic perspective by focusing on the *rapprochement* of practical expertise in water management and mathematical sciences which took place in the period 1550-1625. Although the involvement of late Renaissance mathematicians in projects of water management was far from negligible, this movement was not uniformly spread among European mathematicians. Besides, a comparison between Italy and Holland demonstrates that its features varied according to national traditions and regional identities. Then, by considering the case of Giovan Battista Aleotti, a Renaissance engineer and architect from Ferrara, the first steps of the hydraulic profession will be viewed from a closer vantage-point. It will be argued that the mathematical approach which was slowly developing in the discipline of hydraulics was a token of a new emerging professional group. This group, the so-called "mathematical architects", was an expression of the courtly culture of the Renaissance, of the humanist concern for the recovery of the Vitruvian ideal of architect. However, it was not the architects or engineers who were to shape the new science of hydraulics but a few mathematicians and philosophers who were to gain a repute in the field thanks to their involvement in the controversial context in which Italian Renaissance hydraulics was embedded.

Mathematicians' involvement in hydraulic engineering

As part of a project for a social history of the scientific community of the sixteenth and seventeenth centuries, in the early 1990s the late Professor Richard S. Westfall assembled a database of Western scientists who are included in the *Dictionary of Scientific Biography* (*DSB*). 630 scientists born between 1471 and 1680 are listed in the version of Westfall's catalogue, which I have consulted. They are classified according to a number of headings such as " scientific disciplines ", " sources of support ", " patronage ", &c. What matters in this context is the heading "technological connections ", that is, involvement in projects of practical utility and, within it, the category " hydraulic engineering ". If of these 630 scientists we consider only the mathematicians born between 1525 and 1599 the sample, although reduced to about one sixth (103), is still large enough to have statistical significance[1]. Following Westfall's categories and classifications, I have entered the data of these 103 mathematicians in a table, in which they are classified both according to their involvement in hydraulic engineering and to the geographical area in which they (mainly) lived and developed their careers (see Table 1). Out of 103 European mathematicians, 23 (22.3%) were involved at least once in matters of water management. This percentage rises to more than 50% if we consider Italian mathematicians only[2].

TABLE 1.

European mathematicians in *DSB* born in 1525-1599
Involvement (general) in hydraulic engineering[3]

GEOGRAPHICAL AREA	HYDRAULIC ENG. (Yes)	HYDRAULIC ENG. (No)	TOTAL
Italy	14	11	25
Germany	2	18	20
France	1	15	16
Low Countries	3	12	15
England & Scotland	3	10	13
Other areas	-	14	14
Total	23	80	103

1. By a " mathematician " Westfall meant someone who had made at least one contribution to mathematics or to the physico-mathematical sciences (astronomy, cartography, hydrostatics & hydraulics, mechanics, music, navigation, optics and pneumatics). I accept Westfall's criterion as a provisional starting point. I am aware, as certainly Westfall was, of the pitfalls involved in any classification of the mathematical community of the 16th and 17th centuries. This awareness is, however, no excuse for giving up the tasks of mapping the late Renaissance mathematical community and of studying its social and disciplinary dynamics. As for the time span, I assume that the mathematicians born in 1525-1599 were a good approximation of the group of persons actively engaged in the mathematical sciences in the late Renaissance age.

2. Out of 25 Italian mathematicians, 14 (56%) were at least once involved in water management.

3. Table 1 : in the column *Hydraulic engineering* (Yes) I take into account only those late Renaissance mathematicians who are included in Charles C. Gillispie's *Dictionary of Scientific Biography* (*DSB*) and who, according to Westfall, were involved at least once in water management. " What I want here — wrote Westfall when he was planning his catalogue —, is not Baconian talk about useful knowledge, … , but involvement in actual projects of utility ". The degree and the kind of involvement taken into consideration greatly vary from one individual to another. For this reason I label as " general " this kind of hydraulic involvement.

These are unexpected high figures, at least for the current standards in the history of early modern science and technology, and deserve a closer study. By examining the biographical data assembled by Westfall we realize that part of the mathematicians who have been included by him in the category " hydraulic engineering " (involvement in) neither supervised any important hydraulic scheme, nor were consulted as hydraulic experts in drainage projects and other public waterworks, nor devised engineering plan of canalization and river control. Their inclusion has been due to the writing of books and treatises (Guidobaldo del Monte and Giambattista della Porta), to the design and construction of fountains (Benedetti and Kepler), to the construction of hydraulic pumps and other kinds of water supplies for monasteries and manor-houses (Cavalieri, Desargues and Harriot), to the building and improvement of windmills and waterwheels (Johann Faulhaber and Jan Cornets de Groot), or to other inventions (Napier). In order to appraise the degree of involvement of the mathematicians in the emerging hydraulic profession, this kind of research and activity should not be considered as pertinent to the bulk of late Renaissance hydraulic engineering[4]. If we enter Westfall's data in a new table according to this criterion, the corresponding figures measuring the degree of hydraulic involvement become much lower see *Table* 2)[5]. However, they still suggest the existence of a remarkable hydraulic activity of late Renaissance mathematicians. By assuming that the work of the mathematicians included in the *Dictionary of Scientific Biography* was representative of late Renaissance mathematical activity in the various areas into which Western Europe was divided, we might conclude that in this age about 40% of Italian mathematicians (10 out of 25), 13% of Dutch mathematicians (2 out of 15) and 7.5% of British mathematicians (1 out of 13) were involved in the hydraulic profession[6]. On the European scale this percentage is still greater than 12.5% (13 mathematicians out of 103), again an unexpectedly high figure although almost scaled down by half in comparison with Table 1.

4. In the strict sense this kind of research either belonged to practical mechanics or to scientific hydromechanics. On the other hand, if we were to assess the degree of involvement of mathematicians in the science of waters we would have to consider not only some of the above mentioned characters (e.g. Benedetti and Cavalieri), but also people such as Descartes, Magiotti, Mersenne and Riccioli (all listed in *DSB* and all born in the period 1525-1599). Besides, we would have to evaluate the influence exerted by Renaissance mathematicians such as Buteo, Cardano and Tartaglia, listed in *DSB* but born before the period under consideration.

5. This is not to say that in other geographical areas there was no scientific involvement in water management. Even by limiting the sample to the scientists listed in *DSB*, we may refer for France to J. Buteo, as well as to B. Palissy, who constructed the fountains in the Tuileries. In Flanders it was the Royal Cosmographer and Mathematician, M. Florent van Langren (born in *c.* 1600), who was most active as an hydraulic engineer. If we were to look outside of Westfall's list of 630 scientists, it would be possible to find examples of this kind for other European countries.

6. The biographical data on which tables 1 and 2 are based are those which were available to Westfall and collaborators in the early 1990s. Further historical research may bring to light new hydraulic connections and may lead to a revision of our figures.

TABLE 2.

European mathematicians in *DSB* born in 1525-1599
Involvement (strict) in hydraulic engineering[7]

GEOGRAPHICAL AREA	HYDRAULIC ENG. (Yes)	HYDRAULIC ENG. (No)	TOTAL
Italy	10	15	25
Germany	-	20	20
France	-	16	16
Low Countries	2	13	15
England & Scotland	1	12	13
Other areas	-	14	14
Total	13	90	103

It may be useful to look more closely at the kind of hydraulic activity prac-
tised by these 13 mathematicians[8]. Some of them actually supervised important hydraulic schemes (e.g. Bombelli and Dudley, who respectively worked at
the reclamation of the Val di Chiana and of a swamp between Pisa and
Livorno), or devised one or more engineering projects of canalization and river
control (e.g. Stevin for the Weichsel in Danzig and Veranzio for the Tiber in
Rome). However, the most usual involvement was as technical advisers and
hydraulic experts. Baliani acted as expert for some waterworks in Genoa.
Beeckman was consulted on a plan to rid the harbour of Middleburg of sand-
bars. Cabeo was employed by the Gonzaga in Mantua on hydraulic projects
and gave advice in Ferrara on the regulation of the Po. Castelli supervised the
technical side of a visit to the waters of Ferrara and Bologna. Danti gave
advice on hydraulic matters both in Florence and in Bologna. Galileo was consulted about a flood control project of a river near Florence. Patrizi acted as
hydraulic adviser in Ferrara. Ricci was called to Ferrara to offer an opinion on
the diversion of the river Reno. Wright was the technical expert on the New
River project, a waterway designed to bring water to London. Although the
divide is not always clear-cut, it is apparent that some of them acted mainly as
engineers, others as intellectuals. While Bombelli and Stevin are good examples of the engineering type, the hydraulic involvement of court philosophers
such as Patrizi and Galileo had more to do with the intellectual sphere than
with the work in the field.

From the above mentioned data it seems possible to conclude that in the late
Renaissance age most of mathematicians' involvement in hydraulic engineer-

7. Table 2 : in the column *Hydraulic engineering* (Yes) I take into account only the mathematicians of Table 1 who supervised at least one important hydraulic scheme ; or were consulted at
least once as hydraulic experts in drainage projects and other public waterworks ; or devised at
least one engineering plan of canalization and river control. Given that these criteria are stricter
than those used by Westfall, I label as " strict " this kind of hydraulic involvement.

8. Their names are the following : G. Battista Baliani (1582-1666), I. Beeckman (1588-1637),
R. Bombelli (1526-*ca.* 1572), N. Cabeo (1586-1650), B. Castelli (*ca.* 1577-1643), E. Danti (1536-
1586), R. Dudley (1573-1649), G. Galilei (1564-1642), F. Patrizi (1529-1597), O. Ricci (1540-
1603), S. Stevin (1548-1620), F. Veranzio (1551-1617) and E. Wright (1561-1615).

ing was concentrated in Italy and in the Netherlands. There were, however, important diversities between the hydraulic technologies and cultures of the two areas. Although upper Italy and the Netherlands were the centres of an impressive movement of land reclamation, the demand for new drained land was not evenly echoed in the hydraulic commitment of the mathematicians of the two areas. In the northern Netherlands, land reclamation and water management remained almost exclusively the province of practical men such as Andries Vierlingh and Jan Leeghwater. To enter in the field, mathematicians such as Simon Stevin had to dress in their engineering clothes. To put it another way, the scientific activity of Dutch and Flemish mathematicians was only occasionally stimulated by the movement of land reclamation. Not only that. When this was the case, it was more oriented towards mechanics, both practical and theoretical, than towards hydraulics. It seems not coincidental that Stevin wrote a treatise on drainage windmills, while his contribution to river control is mainly confined to a series of projects entrusted to him outside the Netherlands. In late Renaissance Italy, on the contrary, reclamation and river hydraulics were strictly connected activities and the engineering profession had to face the entrance into the field of some of the most distinguished mathematicians of the age. This movement was only indirectly caused by population growth and the new demand for food supply[9]. In the Po valley, rivers usually were at the borders of districts and states. The control of flooding as well as the drainage of marshlands were not only matters of administrative, economic and technological concern but were also political issues. Towns, districts and states had an interest in the management of the waters in the neighbouring territories and tried to have a voice in it. The administrative and political fragmentation gave room for controversies in which the figure of the technical expert and adviser, usually a mathematical practitioner, was prominent. The longest and the most famous of these controversies was the debate concerning the river Reno. A large number of Italian mathematicians took part, during a period of no less than two centuries, in this discussion[10]. In a period of sustained mathematical development such as the late Renaissance age it seems thus only natural that Benedetto Castelli, a lecturer of mathematics in Pisa and Rome and one of the most intimate friends and collaborators of Galileo, was to profit from the experience gained through his involvement in the Reno controversy so as to build up a new science, a science which aimed to be a substitute for Renaissance hydraulics[11].

9. F. Cazzola, " Il " ritorno alla terra ", *Storia della società italiana : Il tramonto del Rinascimento*, Milano, 1987, 145-163.

10. Five (*i.e.* Cabeo, Castelli, Danti, Patrizi and Ricci) of our ten Italian mathematicians and philosophers involved in hydraulic engineering actively participated in the Reno debate (see Cesare S. Maffioli, " La controversia tra Ferrara e Bologna sulle acque del Reno. L'ingresso dei matematici (1578-1625) ", A. Fiocca (ed.), *Giambattista Aleotti (1546-1636) e gli ingegneri del Rinascimento*, forthcoming).

11. C.S. Maffioli, *Out of Galileo. The science of waters 1628-1718*, Rotterdam, 1994, *passim*.

ALEOTTI AND THE MATHEMATICAL ARCHITECTS

We have now to turn our attention to the other side of the coin, namely to the early attempts at mathematization by the practitioners of hydraulic engineering. This part of the story is still less well known than the former. I shall not attempt in this limited work an exhaustive survey. Instead, I shall focus on a particular professional group, the so-called " mathematical architects ", and on a single character among them, the Ferrara architect and engineer Giovan Battista Aleotti (1546-1636).

In 1598 the state of Ferrara passed under papal control. Apparently, Aleotti's professional activity was scarcely affected by this event. Our character served both Alfonso II of Este and the new governors from Rome, and acted as architect and engineer to the city of Ferrara. His professional commitment concerned a wide range of subjects and activities. Already in 1566, at the age of 20, he levelled and surveyed in the Po delta area.

In 1582 he was entrusted with the division of the reclaimed land in the Polesine of Ferrara. Meanwhile he supervised the construction of some of the Ferrara ramparts and fortifications, and designed water-supplies and fountains as well as devices for theatrical effects and surveying instruments.

From 1598 to 1604 Aleotti was heavily involved in water management schemes and discussions. In his capacity as Ferrara hydraulic expert, he took part in the visit to the waters of the Po delta ordered in 1598 by Clement VIII. Most characteristically, Aleotti accompanied his hydraulic plans and writings with a detailed cartographic survey of the region. His maps, and particularly his great chorography of the state of Ferrara, made him known as one of the best cartographers of the age. Aleotti planned various drainage schemes and investigated the silting up of the Ferrara branches of the river Po.

In about 1600-1601 he supervised the construction of a canal connecting Ferrara to the Po grande, a project executed in order to restore partially the navigation facilities of the town. From 1605 onward Aleotti's architectural commitment prevailed. He designed city gates as well as towers and façades of several churches and *palazzi*, and other civic and religious buildings such as theatres and oratories. His masterpiece, the *Teatro Farnese* at Parma, which was planned and built in his late years, is regarded as a landmark in the history of modern theatre building. Like many other Renaissance architects and engineers, Aleotti tried to present himself both as a practical and as a literary man. His library included hundreds of books and he knew enough Latin to translate and to publish the first Italian version of Hero's *Pneumatica* (*Gli artifitiosi et curiosi moti spiritali di Herrone*, Ferrara 1589).

In 1601 he also published the *Difesa per riparare alla sommersione del Polesine di S. Giorgio e alla rovina dello Stato di Ferrara*, a work in which Aleotti put forward his own remedy for restoring the hydraulic equilibrium of the region. Unfortunately his great treatise on water management, which bears the title *Hidrologia overo Ragionamento della Scienza et Arte dell'acque*, remained unpublished. Most of it was written in 1598-1604, when Aleotti was busy polemizing against the hydraulic experts from Bologna and from Rome. Although in the following three decades the book was taken up several times, in the end it remained incomplete. If published, it would have been the first modern treatise of hydraulics[12].

Late Renaissance architects and engineers were accustomed to apply themselves to a wide variety of subjects and activities. In this respect Aleotti's performances, although very impressive, were not as extraordinary and unique as they might appear to us. The humanist concern for the recovery of ancient wisdom as well as the development of book trade had already stimulated a few mathematicians and engineers to edit and to translate some of the main Greek and Latin mathematical and technological texts. In the neighbouring Venice, an erudite humanist who did not disdain to act as an architect and as a hydraulic engineer, the renowned Fra Giovanni Giocondo from Verona, had published already in 1511 an annotated version of Vitruvius' *De architectura*.

In 1543 it was the turn of a teacher of mathematics, who was to gain a European reputation for his findings in algebra, ballistics and hydrostatics. This was Niccolò Tartaglia from Brescia, who published the first translation in a vernacular language of Euclid's *Elements*. Finally, it was a Venetian surveyor and an outstanding cartographer, Giacomo Gastaldi, who contributed to the 1548 translation into Italian of Ptolemy's *Geographia*.

The mathematical sciences were instrumental in favouring the *rapprochement* of high classical learning and the mechanical arts. This is particularly visible in the vast field of architecture[13]. In the preface of his translation of Euclid, Tartaglia referred to the analogical correspondence of macrocosm and microcosm, and pointed to the Vitruvian architects who tried to proportion their buildings to the similarity of the human body, which embodies the ideal

12. An edition of Aleotti's *Hidrologia* has been planned by the Istituto di Studi Rinascimentali of Ferrara, and is currently prepared under the supervision of M. Rossi. The standard biography of Aleotti is still that by Luigi N. Cittadella, which prefaces his edition of Aleotti's *Dell'interrimento del Po di Ferrara*, Ferrara, 1847. In compiling this biographical sketch I have also made use of A.G. Keller, " Pneumatics, automata and the vacuum in the work of Giambattista Aleotti ", *The British Journal for the History of Science*, III (1967), of A. Fiocca, " Libri d'Architetura et Matematicha " nella biblioteca di G.B. Aleotti ", *Bollettino di Storia delle Scienze Matematiche*, XV (1995), as well as of several articles printed in the proceedings of three one-day meetings held in Ferrara and in Bologna in 1994 and in 1995 [*Giovan Battista Aleotti (1546-1636)*, Seminario di studi, I-II-III sessione, a cura di M. Rossi, Bologna, 1994-1995]. Further biographical information on Aleotti is to be included by A. Fiocca, *op. cit.*, in the volume quoted in note 10.

13. In her study " The contribution of architectural writers to a " scientific " outlook in the fifteenth and sixteenth centuries ", Pamela O. Long has argued that " the ideas disseminated by architectural literature in this period must be conceded an important role in the development of new attitudes which were important to scientific innovation " (*Journal of Medieval and Renaissance Studies*, XV, 2 [1985], 296). Although I wholly share this view, the opposite trend is pivotal in my research, namely the study of late Renaissance developments within the mathematical profession which were to affect practical disciplines such as architecture.

proportions because it has been built by the *Sommo Architettore* himself[14]. This meant first that arithmetic and geometry, the sciences of symmetry and proportions, ought to guide the work of the architect and secondly that architecture was a mathematical art. This message had a deep appeal in the Renaissance age. It was indeed the Vitruvian image " of the human body enclosed in the perfect geometrical form of the circle — as Denis Cosgrove has nicely put it — which was to become the icon of Renaissance humanism "[15].

Aleotti was imbued with this tradition. In the *Hidrologia* he did not hesitate to emphasize that the art of the regulation of waters was part of civil architecture and was subjected to *i precetti di Vitruvio*. The architecture of waters, as Aleotti called it, ought to be the realm of accomplished professionals. Hydraulic practitioners should learn the precepts of civil law as well as mathematical arts and sciences such as drawing and surveying, geography and cartography, natural philosophy, geometry and arithmetic. Those men with a working knowledge of all these Vitruvian subjects were to pass the limits of the ordinary engineers and ought to be called *Architetti Matematici per eccellenza*[16].

CONCLUSIONS

The title of " mathematical architect ", claimed by Aleotti for himself and his peers, hints at the new functions which were performed in the late Renaissance age by a new type of professional men. They no longer considered themselves as mere mechanicians, as surveyors, *periti*, drainage engineers and the like, and appealed to the architect's higher identity in order to enhance their own status. In the early seventeenth century, however, just when Aleotti was still busy writing his book, this vision was already partially superseded by the new scientific professionalism. In the Baroque age the movement of professionalization did not follow the line of the encyclopaedic, universal Renaissance man. Rather, competencies and functions were circumscribed and delimited.

The seventeenth-century entrance of mathematicians and philosophers in the realm of water management was thus to end in a new kind of subordination of practical men to intellectuals. However, this dependence was neither absolute nor uniformly spread throughout Europe. In the Republic of Venice, for example, professors of mathematics of the Padua *Studium* began to have a real influence in the water administration only from the late seventeenth century

---

14. *Euclide Megarense Philosopho : solo introduttore delle scientie mathematice : diligentemente reassettato, et alla integrita ridotto per il degno Professore di tal Scientie, Nicolo Tartalea, Brisciano … ,* Vinegia, 1543, f. 4r.

15. D. Cosgrove, *The Palladian Landscape. Geographical Change And Its Cultural Representations In Sixteenth-Century Italy*, Leicester, 1991, 215.

16. G. Battista Aleotti, " Hidrologia " (Biblioteca Comunale Ariostea, Ferrara, ff. 4r-6r ; the quotes are from the copy kept in the Biblioteca Estense, Modena, Cod. Estense 551, ff. 7v-9v).

onward. In the northern Netherlands, we may trace a similar kind of involvement of the Leyden professors of mathematics only from the late 1720s on[17].

The entrance of mathematicians and philosophers such as Castelli and Cabeo onto the stage of water management implied both a *rapprochement* of the mathematical sciences and practical expertise, as well as a new phase in the professional conflict between mathematically trained people and practical engineers. It was a conflict that the mathematical architects had triggered when they had envisaged distinguishing themselves from the ordinary engineers. They, however, did not foresee that their fight against the theoretical inadequacy of unlearned, if not totally uneducated, engineers was to open the doors to a new class of competitors. They were university professors or the like, who were better trained in the mathematical sciences and better equipped with those rhetorical tools which were a fundamental resource when intervening in the controversies related to the " architecture of waters ".

17. C.S. Maffioli, *Out of Galileo...*, *op. cit.*, see note 11 above, 276-279 ; *Idem*, " Italian hydraulics and experimental physics in 18th century Holland ", in C.S. Maffioli, L.C. Palm (eds), *Italian scientists in the Low Countries in the 17th and 18th centuries*, Amsterdam, 1989, 253-254.

The Mathematician as Engineer in the Seventeenth Century : Leibniz and Engineering Hydraulics

James G. O'Hara

Introduction : Leibniz and the Harz Project

In the years between 1679 and 1885 Leibniz spent the equivalent of more than three years working in the Harz mountains developing a project in which wind and water power was to be exploited for draining the mines and improving ore production. Besides efforts to improve the efficiency of wind mills, he developed an elaborate scheme for water-management in the mining district around the towns of Clausthal and Zellerfeld with reservoirs, channels and pumping stations. In effect he tried to develop a system similar to twentieth-century pump storage schemes that would provide a constant supply of water power to drive the pumping machinery in the mines. One interesting but neglected aspect of Leibniz's activity in the Harz mountains was the experience and knowledge he gained in hydraulic engineering particularly in the construction of channels and reservoirs[1]. Following the abandonment of the Harz project in 1685 Leibniz became historian or historiographer of the duchy of Brunswick-Lüneburg and in this capacity undertook, between November 1687 and June 1690, a research tour through southern Germany, Bohemia, Austria and Italy. In 1689 and 1690 he came in contact with a number of hydraulics experts in Italy where there was a long tradition in this field. The discussions which took place are reflected in Leibniz's correspondence during and following the Italian journey.

Hydraulics had already been extensively applied in Italy in medieval times and during the Renaissance, particularly for the control of water and for land reclamation. In the sixteenth and seventeenth centuries it found increasing use in order to improve the flow of natural water, to avoid or reduce flood damage,

---

1. J. Gottschalk, " Theorie und Praxis bei Leibniz im Bereich der Technik, dargestellt am Beispiel der Wasserwirtschaft des Oberharzer Bergbaues ", *Studia Leibnitiana*, Suppl., 22 (1982), 46-57.

to supply water to towns and cities, to promote commercial traffic along exca-
vated canals, to keep ports working, and to improve the operation of mills
along rivers[2]. The problems confronting Italian engineers were of course quite
different in scale and kind from those that Leibniz had faced in his Harz
project. However these different applications of hydraulics did have in common that they were all to open-channel flow with the same concepts and flow
principles applying. Accordingly Leibniz's meeting with the Italian mathematicians and engineers in 1690 amounted to a meeting of experts in applied
hydraulics.

THE GALILEAN TRADITION IN HYDRAULICS

Galileo's discovery of the law of falling bodies and of the parabolic trajec-
tory of a projectile provided the foundation for a description of fluid effluxions
and jets. However in hydraulics Galileo's own contributions were insignificant
compared to those of his disciples, the most important of whom included
Benedetto Castelli and Evangelista Torricelli. Castelli devoted much of the
later part of his life to the study of hydraulics and in 1628 he published at
Rome the first book of his principal work *Della misura delle acque correnti*.
The work was enlarged and a second Roman edition published in 1639. A third
edition containing a second book, written by Castelli in 1642, was edited by
his disciples and published posthumously at Bologna in 1660[3]. Castelli's work
mainly illustrates and demonstrates the truth of a single principle, namely the
continuity principle of hydraulics. The principle is founded, firstly, upon a definition of rate of flow as the volume passing through a cross-section in unit
time and, secondly, upon the principle of the conservation of volumes which,
for steady flow, requires the equality of rates of flow in all cross-sections of a
channel or pipe. In practical application the continuity equation is used, which
states that the rate of flow equals cross-sectional area times velocity and, in
addition, that cross-sectional area and velocity are inversely proportional. The
edition of 1660 contains, besides Castelli's original exposition of the continuity
principle and corresponding geometrical demonstrations, a second book with
further definitions, propositions and theorems on mensurations. Furthermore
specific considerations concerning the lagoon of Venice, the draining of the
Pontine Fens and of the territories of Bologna, Ferrara and Romagna were
appended to the work.

Following the death of Castelli in 1643, Torricelli emerged as the main vir-
tuoso of the Galilean school until his premature death in 1647[4]. His publica-

---

2. C. Maccagni, " Galileo, Castelli, Torricelli and others. The Italian school of hydraulics in the
16th and 17th centuries ", in G. Garbrecht (ed.), *Hydraulics and hydraulic research. A historical
review*, Rotterdam, Boston, 1987, 81-88.

3. B. Castelli, *Della misura dell' acque correnti*, Bologna, 1660.

4. E. Levi, " Evangelista Torricelli ", in G. Garbrecht (ed.), *loc. cit.* (n. 2), 93-102.

tion of the efflux law in 1644 provided the basis for the study of flow through orifices and slots. A further generation of hydraulics experts emerged late the seventeenth century, the most important of whom was Domenico Guglielmini. The latter was author of two important works namely *De aquarum fluentium mensura nova methodo inquisita* (1690) and *Della natura dei fiumi* (1697). While the later treatise on the theory of rivers was undoubtedly Guglielmini's magnum opus that helped found fluvial hydraulics on a new basis, it is the earlier work on the mensuration of running waters, written while he was professor of mathematics at Bologna, that is of interest here[5]. In this work we see a continuation of approaches adopted by Castelli and Torricelli[6]. All in all, Castelli, Torricelli and Guglielmini represent the outstanding figures of seventeenth-century Italian hydraulics and were in the mainstream of a school or tradition of hydraulics which was probably the most important in Europe between the Scientific Revolution and the Enlightenment[7]. It need not surprise therefore that Leibniz was attracted to their work.

FLOW AND DISCHARGE PRINCIPLES

One specific problem that particularly interested Leibniz at the time of his Italian journey must be elaborated on at this point. Castelli's treatise on mensuration represented the first step towards a mathematical treatment of fluvial hydraulics[8]. As indicated already, much of the work is taken up with an exposition of the continuity principle and in this way Castelli established one of the fundamental relations required for the understanding, design and management of the flow of water in rivers and canals. He did however leave some problems unsolved which he had discussed with Galileo and the mathematician Bonaventura Cavalieri. One such problem concerning the different velocities observed at different depths of a stream was treated for example in a letter sent by Cavalieri to Castelli on 11 January 1642[9]. This problem continued to be discussed by the Galileans and the posthumous edition of 1660 contains as the second proposition of the second book the statement that in a stream passing through a regulating device of rectangular cross-section the velocity is proportional to depth. By restricting the investigation to flow through such a rectangular cross-section, Castelli was able to introduce a considerable simplification and adopt a method of analysis in which the velocity of the stream through the

5. D. Guglielmini, *Aquarum fluentium mensura nova methodo inquisita*, Bologna, 1690.

6. D. Barsanti, " La scuola idraulica Galileiana operante in Toscana ", *Bollettino Storico*, Pisano, 58 (1989), 83-129.

7. C.S. Maffioli, " Guglielmini vs. Papin (1691-1697). Science in Bologna at the end of the 17th century ", *Janus*, 71 (1984), 63-105 ; C.S. Maffioli, *Out of Galileo. The Science of Waters 1628-1718*, Rotterdam, 1994.

8. J.C.I. Dooge, " Historical development of concepts in open channel flow ", in G. Garbrecht (ed.), *loc. cit.* (n. 2), 205-229.

9. M. Bucianti (ed.), *Benedetto Castelli Carteggio*, Florence, 1988, 243f.

section and the depth of flow were the only variables. For the variable depth of flow he used the term *altezza viva* which Thomas Salusbury translated in 1661 as " quick height ". In this translation Castelli's proposition is enunciated as follows : " If a River moving with such a certain velocitie through its Regulator, shall have a given quick height, and afterwards by new water shall increase to be double, it shall also increase double in velocitie "[10]. In the accompanying demonstration the fluid in the rectangular cross-section was conceived as divided up into a series of layers all of which are projected with a common velocity. Each layer has, however, in addition to its own velocity also that of the layer immediately below it and thus a linear velocity distribution increasing from the bottom to the top layer is inferred.

In 1644, the year following Castelli's death, Torricelli published his celebrated law according to which the velocity of the efflux of water through an orifice in a vessel, in which the level is kept constant, is proportional to the square root of the height of water above the orifice. In a section of the work *De motu gravium naturaliter descendentium et proiectorum libri duo* entitled *De motu aquarum* he demonstrated this law as well as a parabolic distribution of efflux velocity from the top to the bottom of the vessel *i.e.* with the zero-velocity vertex at the water surface[11].

The starting point of Guglielmini's researches, described in his treatise *De aquarum fluentium mensura nova methodo inquisita*, of 1690 are results obtained by Castelli and Torricelli. In fact Guglielmini's intention in writing the work was to overcome the shortcomings of Castelli's treatise and in particular his failure to demonstrate conclusively a distribution of velocity with depth. The work was divided into six books or chapters the first three of which were published in 1690. The remaining three books that make up the second part were published in 1691. Having given an exposition of Castelli's continuity principle in the first book, Guglielmini treats Torricelli's efflux law in the first proposition of the second book ; this forms the basis for the distribution of velocities in flow along sloping and horizontal open channels which are conceived as straight oblong containers with plane sides and bottoms. In the second proposition of the third book he applies the efflux law to the flow of water along such a horizontal open channel. The velocity of flow at a given depth is considered to be the same as that from an orifice in a container under a head equivalent to this depth, and accordingly the velocity at various depths is proportional to the square root of the distance below the surface. Guglielmini thus postulated that the velocity distribution in a stream must be parabolic with the zero-velocity vertex near the water surface ; it is represented by the curve

10. B. Castelli, *op. cit.* (n. 3), 82 ; T. Salusbury, " On the mensuration of running waters ... by Don Benedetto Castelli ", *Mathematical Collections and Translations*, London, 1661 (facs. London, Los Angeles, 1967.

11. G. Loria, G. Vassura (eds), *Opere di Evangelista Torricelli*, vol. II, Faenza, 1919 and 1944, 190, 4 vols.

KIHEC in the following (Fig. 4) that is taken from an account of Guglielmini's book given by Leibniz in the Leipzig journal *Acta eruditorum* in 1691[12].

FIGURE 4.

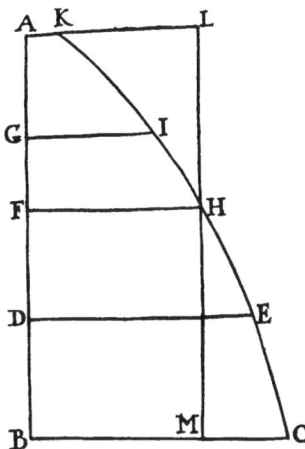

Guglielmini was aware that the parabolic distribution of velocity he was proposing would not apply in real rivers and streams and he suggested it would only be obtained with a perfect fluid. But there is a more fundamental flaw in this application of the efflux law to obtain a velocity distribution in streams which would render it invalid even in such a hypothetical perfect fluid. Torricelli's law is based on Galileo's law of falling bodies and applies only to flow through an orifice or slot into the free atmosphere. An element of fluid in a stream, on the other hand, is subject not only to the influence of gravity but also to pressure from the surrounding fluid. Nevertheless this erroneous application of the efflux law and the resulting velocity distribution — the so-called *scala fallacy*[13] — enjoyed considerable support in the late seventeenth- and early eighteenth-century Italy[14]. Guglielmini did, nevertheless, give a second and this time valid application of Torricelli's efflux law, namely to flow along sloping channels. Thus the second proposition of the second book states that the velocity of water passing through any section of a sloping channel is that which it would have if it issued from an orifice equal to this section and under a head equal to the vertical distance between the section and the horizontal

12. Anon. [G.W. Leibniz], " Aquarum fluentium mensura nova methodo inquisita autore Dominico Gulielmino M.D. ", *Acta eruditorum* (1691), 72-75.

13. S. Leliavsky Bey, " Historic development of the theory of the flow of water in canals and rivers ", *The Engineer*, 191 (1951), 466-567, 601-603. *Cf.* 466.

14. G. Cocchi, " Le discipline idrauliche nella storia della scienza " (Hydraulics in the history of science), *Alma Mater Studiorum, Revista scientifica dell' Università di Bologna*, III (2) (1990), 65-102, *cf.* 93f.

passing through the point of commencement of the channel. From this it follows that the velocity at any point along the sloping channel, F or B for example along the channel AB in (Fig. 5) of Leibniz's 1691 review, is proportional to the square root of the corresponding vertical distance below the horizontal plane passing through the beginning of the channel[15].

FIGURE 5.

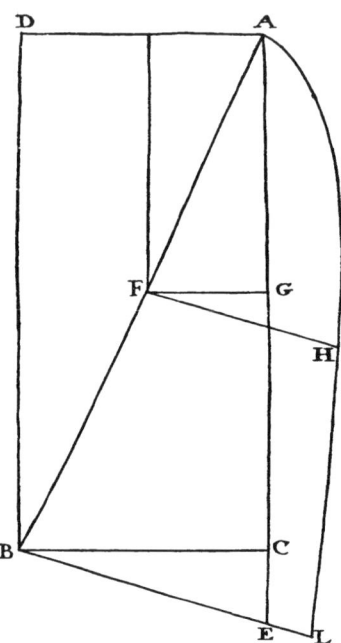

LEIBNIZ'S MEETING WITH THE ITALIAN HYDRAULICIANS

The only member of the original Galilean school still living at the time of Leibniz's Italian journey was Vincenzo Viviani, a mathematician with interests in hydraulics. However when Leibniz met him in Florence they appear not to have discussed questions of hydraulics, at least as far as we are able to glean from remarks in Leibniz's correspondence for example with Christiaan Huygens on 25 July 1690[16]. The most fruitful meeting as regards hydraulics was undoubtedly that with Guglielmini whom Leibniz met in Bologna in late December 1689. Guglielmini reported on 8 January 1690 to Antonio Magliabechi, the influential librarian of the Grand Duke of Tuscany, on his meeting

15. *Cf.* Anon. [G.W. Leibniz], *op. cit.* (n. 12).

16. *Cf.* G.W. Leibniz, *Gottfried Wilhelm Leibniz. Sämtliche Schriften und Briefe*, III, 4, Berlin, 1923 -, 532-538, 29+ vols in 7 series.

with Leibniz[17] and a correspondence between Guglielmini and Leibniz was initiated.

For the years 1690 and 1691 we find however no detailed discussion of problems of hydraulics[18]. In fact the most detailed information on the meeting with Guglielmini and others interested in hydraulics is found in Leibniz's correspondence with the Modenese physician Bernardino Ramazzini whom he had met at the end of December 1689 and with whom he had several meetings during his five-week stay in Modena. Throughout the decade 1690-1700 correspondence was kept up and ten letters from Ramazzini and six from Leibniz are extant. Ramazzini himself had interests in hydraulics and was about to publish a work *De fontium Mutinensium admiranda scaturigine* (1691) about the underground streams of Modena making use for the first time of the piezometer and the idea of piezometric head for flow in closed conduits[19]. Ramazzini also appears to have indicated to Leibniz his intention of undertaking experiments to observe an increase in the depth of flow or surface elevation caused by an obstacle placed in an open channel. Thus in a letter from Venice on 25 February 1690 we find Leibniz requesting the communication of such results[20]. The same letter also contains the intelligence that Guglielmini had confided to him that he had a work in hand in which he hoped to establish various laws of fluvial motion on a sounder foundation than Castelli had been able to do.

Replying on 15 April Ramazzini provides details of an investigation carried out by a certain Modenese expert, whom Leibniz had likewise met during his stay in the town, namely Giovanni Battista Boccabadati[21]. The latter was planning a publication on applied hydraulics based on field work carried out on two tributaries of the Po, namely the Scoltenna, which is the Latin name of a section of the Panaro, and the Gabelus or Sécchia. In times of considerable precipitation and unfavourable winds these rivers, which flow on each side of Modena, were apt to burst their banks and to flood the entire region. Ramazzini informs Leibniz that on the recent occasion of such flooding, on 2 and 3 April 1690, Boccabadati had made observations along the Gabelus which crosses the Via Aemilia between Modena and Reggio and then follows a winding course leading to the Po. There Boccabadati claimed to have observed a considerable increase in the water level over a relatively short distance at a point where

17. G.W. Leibniz, *Gottfried Wilhelm Leibniz. Sämtliche Schriften und Briefe, op. cit.*, I, 5, 684-685.

18. *Cf.* M. Cavazza, " La corrispondenza inedita tra G.W. Leibniz, Domenico Guglielmini e Gabriele Manfredi ", *Studi e Memorie per la Storia dell' Università di Bologna (nuova serie)*, 6 (1987), 51-80 ; *Cf.* A. Robinet, " G.W. Leibniz et la république des lettres de Bologne ", *Studi e Memorie per la Storia dell' Università di Bologna (nuova serie)*, 6 (1987), 3-49 ; *G.W. Leibniz, Iter Italicum*, Florence, 1988, 296-309.

19. *Cf.* C.S. Maffioli, *op. cit.* (n. 7, 1994), 215-218.

20. *Cf.* G.W. Leibniz, *op. cit.* (n. 16), III, 4, 466-468.

21. *Idem*, III, 4, 499-500.

flood water rapidly impinged against a curved embankment and was accordingly diverted from its natural path. Boccabadati's interpretation of his observation was that the water above the embankment was much higher than below as a consequence of the fact that at impact eddy currents were generated which in turn slowed down the oncoming water ; on the other hand the water below the embankment being free of any such impediment could flow away easily.

Boccabadati's observation can of course be explained in terms of Castelli's continuity relation. Under conditions of steady uniform flow such a diversion of the stream leads to retardation and therefore a swelling of the water above the diverting structure, whereas below it the water is able to run off rapidly and the level therefore drops. Leibniz, replying to Ramazzini from Hanover on 16 July 1690, surely had Castelli's principle in mind when he remarked that Boccabadati had indeed most correctly observed a difference in water level[22]. Leibniz then turns to the velocity distribution with depth. It would be interesting to investigate and establish the amount by which the velocity increases with increasing depth, he suggests. Castelli had proposed a theorem for this distribution but had been unable to demonstrate it satisfactorily. Leibniz, for his part, doubted that it would be possible to establish a constant rule at all. In the letter that was sent to Ramazzini, and even more explicitly in a variant reading in the draft of the letter, Leibniz extended this problem of finding a velocity distribution with depth to that of finding the velocity at a point downstream from the place of increase in depth of flow.

The problem is stated as follows : if AB (shown as a straight line in a marginal diagram) represents a long stretch of a river and if at the point A, and only at this point, a change in depth of flow takes place for any reason whatsoever, find the increase in velocity at a point near B at a considerable distance downstream from A. Leibniz suggests that the increase in velocity observed would be almost negligible although the increase in depth of flow might be very considerable. In the dispatched letter he states again that Guglielmini had told him that he had a book in hand about this matter and would provide a different rule to that given by Castelli. But Leibniz remained sceptical about the possibility of obtaining such a rule at all.

Ramazzini took several months to answer Leibniz and Rudolf Christian von Bodenhausen, a German preceptor resident in Italy, now emerged as Leibniz's principal informant on the progress of Guglielmini's work. The first news of the impending publication of Guglielmini's book came in a letter from Magliabechi in Florence dated 5 August[23] but in the end it was Bodenhausen who, also writing from Florence on 16 September 1690, informed him that Guglielmini's book had at last been published[24]. At the same time Bodenhausen

22. G.W. Leibniz, *op. cit.* (n. 16), III, 4, 531-532.
23. G.W. Leibniz, *op. cit.* (n. 16), I, 5, 642.
24. G.W. Leibniz, *op. cit.* (n. 16), III, 4, 554-558.

expresses criticism of Guglielmini's work, the tenor of which is that it consisted essentially of variations on a single theme that had been unnecessarily inflated into a book. By early November Leibniz was most anxious to see Guglielmini's opus. Writing to Magliabechi on 5 November he expresses his strong desire to see Guglielmini's little book and complains about the difficulty of obtaining such publications in Hanover[25]. It would be a matter of great importance, he says, if Guglielmini could supply a demonstration of the theorem that Castelli had tried in vain to verify. He did not doubt the capability of Guglielmini and had great expectations for the work. Enclosed with this letter to Magliabechi was another to Bodenhausen, also dated 5 November 1690[26], in which Leibniz asks his correspondent for a report on the main propositions and the underlying principles of Guglielmini's treatise.

Shortly after writing these letters Leibniz at last obtained a review copy for the *Acta Eruditorum* from the editor Otto Mencke in Leipzig with an accompanying letter dated 7 November 1690[27]. In his next letter to Bodenhausen on the 17 November[28] we find Leibniz's first reaction on reading the work. Singling out Guglielmini's parabolic distribution of velocity with depth Leibniz expresses the conviction that what the author had written in his first three books could not apply in real rivers and other large-scale works, as indeed the author himself admitted, for the incremental change of velocity, which was known to be in the sub duplicate ratio of the depth below the surface, would be totally altered in the flowing and dashing of the water. Hence this velocity distribution would not apply in practice. It is interesting to notice that Leibniz did not reject the Torricellian velocity distribution as such ; his sole objection was simply that it would not apply in practice.

Replying from Florence on 19 January 1691[29] Bodenhausen expresses his conviction that Guglielmini, though ambitious, did not have the capability to carry out the task he had set himself and once again repeats and elaborates on his earlier criticism. By this time however Leibniz was in a position to form his own independent judgement. The February 1691 number of the *Acta Eruditorum* included an informative account of Guglielmini's first three books either by Leibniz or adapted from the report he sent to the editors. The reviewer here desisted from any criticism or commentary, except for repeating Guglielmini's own admission that the results would not apply to real rivers and streams. Nonetheless, it was on the basis of this account that Denis Papin was moved to write a critique of Guglielmini's work that was published in the May 1691 number of the *Acta eruditorum*. This in turn led Guglielmini to publish a

25. G.W. Leibniz, *op. cit.* (n. 16), I, 6, 280.

26. G.W. Leibniz, *op. cit.* (n. 16), III, 4, 626-633.

27. G.W. Leibniz, *op. cit.* (n. 16), I, 6, 284-285.

28. G.W. Leibniz, *op. cit.* (n. 16), III, 4, 652-653.

29. *Cf.* ms letter, Niedersächs. Landesbibl. Hannover, Lbr. 79, 45-48 (Bodenhausen-Leibniz, 19.Jan.1691).

rejoinder in 1692, *Epistolae duae hydrostaticae*, in the form of two open letters addressed to Leibniz and Magliabechi, respectively[30]. A commentary by Leibniz on the first part of Guglielmini's *Aquarum fluentium mensura* is found in a letter to Bodenhausen written on 23 March 1691[31]. Once again it is the parabola of velocity distribution that attracts Leibniz's comment. As soon as one takes account of the resistance between the fluid and the bottom and sides of the channel, as well as between the fluid particles themselves, all of Guglielmini's demonstrations collapse, Leibniz suggests, and furthermore just as a river, that has been flowing for some time, will have spent most of its initial impetus received from gravity, so too Guglielmini's demonstrations regarding the parabola, that are likewise all based on a continuation of an impressed force of gravity, are worthless. Thus Leibniz's criticism is twofold. In the first place the whole of Guglielmini's work is only valid for a perfect non-viscous fluid, free of friction with the bottom and sides. Secondly, he considers that for the velocity distribution proposed by Guglielmini to be valid, a constant head would have to be maintained. This is indeed correct but of course there is a further condition necessary for the Torricellian distribution to be applicable, namely the pressure or piezometric head must become zero as is only achieved in discharge through an orifice or slot into the free atmosphere. Leibniz has understood that there is a fundamental flaw in the application but is unable to identify completely the root of the evil. Apart from these shortcomings Leibniz had found Guglielmini's work quite to his liking but thought the author would do well to apply himself to the study of physical principles.

Finally, though not related to the encounter with the Italians, we find in Leibniz's comments on mathematical or physical questions occasional evidence of the influence of his engineering experience. Thus writing to Christiaan Huygens, on 27 May 1691 Leibniz expresses dissatisfaction with Pierre Varignon's theory of gravity, as published in the *Nouvelles conjectures sur la pesanteur* of 1690, and he suggests that it were as if Varignon had proposed that a stream having a constant rapidity or velocity must have a force or head proportional to its length[32]. Leibniz insists that in his view the force or head was manifest solely at a specific point where the stream acted and was unrelated to the river's length.

CONCLUSIONS

The information obtained from Leibniz's correspondence in 1690 and 1691 allows a number of conclusions to be drawn. His meeting with Guglielmini at the turn of the year 1689-1690 seems particularly significant and a possible

---

30. *Cf.* C.S. Maffioli, *op. cit.* (n. 7, 1994), 228-236.

31. *Cf.* ms letter, Niedersächs. Landesbibl. Hannover, Lbr. 79, 43f. And 57 (Leibniz-Bodenhausen, 23 March 1691).

32. *Cf.* C. Huygens, *Oeuvres complètes de Christiaan Huygens*, X, The Hague, 1888-1950, 99-100, 22 vols.

influence of their discussions on the appearance of Guglielmini's book cannot be ruled out. The dedication to the senators of Bologna is dated August 1690, little more than six months after the meeting with Leibniz and it is striking that Guglielmini sets out in the work to give answers to almost the same questions that Leibniz was asking. It is not at all unlikely that the questions found in Leibniz's letters to Ramazzini had already been posed directly to Guglielmini. Leibniz's main interest was to have a demonstration of Castelli's theorem which suggested a linear distribution of the velocity from the bottom to the top of a stream. He also hoped to learn of results of investigations on the flow of water over an obstructing structure in a stream and of the velocity downstream from a point of sudden increase in depth of flow.

Leibniz found at best only partial satisfaction in Guglielmini's work. He quickly realized that the propositions and demonstrations of Guglielmini would not be applicable in real rivers and streams and would hold only for a perfect fluid. He failed however to realize that there was a fundamental mistake in Guglielmini's application of Torricelli's efflux law to obtain a velocity distribution in a stream. Leibniz questions are general and he does not specify, for example, a channel contraction causing an increase in depth of flow or real structures like weirs and sluices but it is tempting to suppose that he had such in mind.

We must remember however that in 1690 only a first approach would have been possible to such gravity phenomena and another generation was to pass before the first formulae were obtained.

In conclusion a whig interpretation of these events will be ventured ; that is, from the vantage point of our present knowledge, the state of development of flow principles and concepts at the time of Leibniz's meeting with the Italian hydraulicians will be considered. The type of problem discussed in Leibniz correspondence with Ramazzini relates to steady flow in open channels with zones of transition between sections of uniform flow. For the analysis of this type of open-channel transition two relationships or equations are required namely the continuity and Bernoulli relationships respectively. As for the continuity relationship we have seen that knowledge of it had been widely diffused through Castelli's work and that Leibniz was well acquainted with it.

The Bernoulli equation, on the other hand, which for open-channel flow expresses a constancy of total head or the sum of velocity head, depth of flow and channel elevation along a streamline, was a development of the 18th century and later[33]. We can however trace the origins of its underlying concepts to the late 17th century.

33. See I. Szabó, " Über die sog. " Bernoullische Gleichung " der Hydromechanik. Die Stromfadentheorie Daniel und Johann Bernoullis ", *Technikgeschichte*, 37 (1970), 27-64 ; C. Thirriot, " L'Hydraulique au fil de l'eau et des ans à travers les XVIIe et XVIIIe siècles ", in G. Garbrecht (ed.), *loc. cit.* (n. 2), 117-143 ; D. Vischer, " Daniel Bernoulli and Leonhard Euler ; the advent of hydromechanics ", in G. Garbrecht (ed.), *loc. cit.* (n. 2), 145-156.

The efflux law in its present-day formulation $v = \sqrt{2gh}$ is easily transformed into the form $h = v^2/2g$ which is the expression for velocity head. It is therefore not surprising that Guglielmini's application of Torricelli's law has been interpreted by twentieth-century authors as the first step towards the Bernoulli equation. Thus for example Hunter Rouse and Simon Ince, in their *History of hydraulics* of 1963, have stressed that Guglielmini's application of the efflux law to flow conditions at a sluice gate or at the inlet of a sloping channel represents the first step towards the present-day graphical representation of the total head and its component parts[34]. He correctly reasoned, they point out, that the velocity at the free surface in the zone of non-uniformity varied with the square root of the distance below the original surface level. A closer study of the writings and correspondence of Leibniz and Guglielmini, while largely confirming this understanding, does allow however a more detailed appraisal of certain specific points. One such consideration relates to Castelli's *altezza viva* or " quick height " and the corresponding term *altitudo viva* used by Guglielmini, the importance of which was also emphasized by Leibniz in his review of February 1691. This quantity represents a variable depth of flow in the sense of the Bernoulli equation. We have seen that Leibniz wanted to generalize Castelli's theorem and formulated a problem of finding the velocity at a point downstream from an increase in depth of flow. Here we have perhaps a first approach to the concept of velocity head since, although Leibniz contemplates only the variation of velocity and not of the square of velocity, he does in fact treat velocity as the key parameter. Moreover, in his remark to Huygens concerning the force or head of a river being manifest at a specific point we find a way of thinking that derives from his engineering experience.

Furthermore, in Leibniz's remark to Bodenhausen that a river will continue to flow only if the force or impetus of gravity continues to operate as well as in his expectation that the velocity increase downstream from the point of increase in depth of flow would be inconsiderable, we have an implicit understanding of friction phenomena and of the dissipation of energy. In this context it may also be recalled that, in terms of our present-day formalism of hydraulics, the product of velocity head ($v^2/2g$) and specific weight, or the weight per unit volume, yields the kinetic energy ($1/2\ mv^2$) per unit volume of the fluid and that the origin of this concept in physics can be traced back to Leibniz's proposal, in his March 1686 article *Brevis demonstratio* in the *Acta eruditorum*, that mv^2 (rather than mv) is the quantity that is conserved in nature[35]. It is also significant that Leibniz had arrived at the square of the velocity from a consideration of Galileo's law of falling bodies which was in turn also central to Torricelli's and Guglielmini's approach.

34. H. Rouse, S. Ince, *History of hydraulics*, New York, 1963, *cf.* 69.

35. G.W. Leibniz, " Brevis demonstratio erroris memorabilis Cartesii et aliorum circa legem naturae ", *Acta eruditorum* (1686), 161-163.

The principal, though tentative, conclusion of this paper is therefore that the meeting of the ideas of Leibniz and Guglielmini in 1690 led to a kind of synthesis out of which developed early notions of a concept of head subdivided into constituent parts as well as of an energy principle in hydraulics. Beyond this we have here a meeting of two mathematicians whose common interest derives from their respective engineering experience.

ACKNOWLEDGEMENT

I am grateful to Professor James Dooge of University College Dublin for his comments on an earlier version of this paper.

Models at the Beginning of Modern Civil Engineering: Development of the Concepts and Three Examples

Bruno MEYER

Modern civil engineering dates back to the Europe of the 18th century. It is closely bound up with the beginning industrialization and creates a new relation to science and research. This contribution shows that civil engineering has been crucially influenced by scientific models and that provides some good examples of how the concept of model has been developing.

Scientific Models and the Rise of a New Profession

The most important feature of modern civil engineering is a specific competence which has led up to the birth of a new profession. Since then, its representatives have called themselves civil engineers. They created new professional groups within the construction industry as well as within the economy and public administration. And they express their new position through professional organizations, separate disciplines in the higher education, and their own technological periodicals.

At the beginning and depending on the country, however, they differed widely in their professional roots, with professions like carpenters, stonemasons, millwrights, metal workers and tool makers as well as military officiers and fortification engineers being known.

The works of civil engineering are manifold, such as roads, bridges, dams and tunnels as well as harbours, waterways, river control systems and coastal protection. Moreover the works for water supply and sewage treatment of the cities should be mentioned, as well as the works for irrigation and drainage in rural areas. All together they are often called " engineering works " — a term which is sometimes used synonymously with " public works ".

Such works already existed in antiquity as they represent the oldest elements of any civilization.

Today they are usually ascribed to civil engineering too. Since the 18th century, however, their variety has been increasing tremendously. Under this presupposition — and only under this one — it is justified to speak of " modern " civil engineering.

What is new about modern civil engineering ? — This question is usually answered linguistically. It is said that the first civil engineers wanted to differ from military engineers, and that they wanted to emphasize their non-military stance. This explanation is too simple and has to be modified. In those times new know-how and knowlegde were developing. They were both evolving together with the requirements of the western civilization which can be illustrated by the design of technical facilities. The new management structures, however, have to be added too. These forms are an essential part of the engineers' professional self-awareness and of their social standing.

The first criterion of this new profession is the fact that its know-how and knowledge stand in a new relation to science. Traditional designing used physical models and outlines which were elaborated into working drawings. In the 18th century such models were already widespread to explain scientific phenomena (fig. 1).

Within technology, however, such models are used to explain the purpose they have to serve. So the engineer is able to communicate his idea of what and how something has to be implemented. From those days it is reported that tests on the basis of models had already been instituted[1].

This method of designing has been enlarged since the 18th century. An *additional theoretical level* is inserted between building and project. In retrospect it may be said that, on this level, the engineers have developed ideas which are today called scientific models — in retrospect — because at that time the concept of model did not comprise anything else than the physical model, *i.e.* the small-scale copy of something to be painted, sculptured or built. Since then traditional design by sketches and models has been interacting with the models on this new theoretical level.

According to the present definition, a scientific model is " a familiar structure or mechanism used as an analogy to interpret a natural phenomenon. Scientific models are employed to develop new theories, to modify existing theories or to give them new applications, or to render theories more intelligible[2]. They have to include the elements to be investigated, and the relation between model and object has to be clarified. Two examples from the 18th century will illustrate these characteristics[3].

1. E.S. Ferguson, *Das innere Auge. Von der Kunst des Ingenieurs*, aus dem Amerikanischen von A. Ehlers, Basel, 1993, 134-140.

2. *The New Encyclopaedia Britannica, Micropaedia,* vol. VI, 15th ed., Chicago, 1974, 958 ; see also *Grand Dictionnaire Encyclopédique Larousse*, tome 7, Paris, 1984, 7005-7007.

3. Both of them are well known and cited in technical publications frequently.

DEVELOPMENT OF STRUCTURAL MODELS

In 1743 three men of learning published an expertise about the bad state of St. Peter's cupola in Rome. Commissioned by the owner, Pope Benedict XIV, they had tried to explain the causes of the alarming cracks. Based on their analysis, they made proposals for structural repairs. They recommended adding further tie rings in order to hold the whole structure together. Yet the three were neither practitioners nor building experts, but theologians and mathematicians. The explanation they gave on the cause of the damage was rejected immediately. Their proposals, however, were accepted at the same time. In 1743 and 1744 an architect installed five other tie rings[4].

Historians of civil engineering call this report a turning-point. They even call it the birth of modern civil engineering[5]. In 1949 Hans Straub wrote : " Its importance lies in the fact that, contrary to all tradition and routine, the stability survey of a structure has been based, not on empiric rules and statical feeling, but on science and research. Its importance lies, furthermore, in the new approach to the problem which is, for the first time, treated as a problem of quantitative statics in the modern sense : the dimensions of a building element (the tie ring) are to be determined directly, by calculation "[6]. The first part of this statement is certainly correct. The second one, however, may lead to the erroneous idea that questions of statics are merely quantitative ones which can be answered by calculation. It may easily be forgotten that they are based on a process of modeling including several levels of abstractions. These statement can be illustrated by the above mentioned report on the bad state of the cupola.

The report itself is structured in a way that is still in common use today. At first the building and the discernible damage are drawn to scale. Then the possible causes are discussed and the existing measures are proved insufficient. Finally a calculated proposal for repair is presented. What matters, though, are the several abstractions made by the three men of learning. It is these abstractions that led to scientific models and gave an essential impulse to enlarging the concept of models in civil engineering.

In the plate added to the report of 1743 (fig. 2) the abstractions can be seen in four steps : the building, the structure, the structural model and the calculation. On the left hand side there is the cupola with its cracks. On the right hand side there are drawn only four arches which converge in a horizontal arch and carry the lantern. Above that, the experts outlined a mechanism of deflection and calculated the resulting horizontal forces to determine the number of tie

4. H. Straub, *Die Geschichte der Bauingenieurkunst*, 4, erw. Aufl., Basel, 1992, 154-162 (1. Aufl., Basel, 1949 ; 1st engl. ed. : H. Straub, *A History of Civil Engineering*, Manchester, 1952).

5. D. Conrad, Th. Hänseroth, " Die 'Geburtsstunde des modernen Bauingenieurwesens' vor 250 Jahren und ihre Vorgeschichte ", *Bautechnik*, 70 (1993), 176-180.

6. H. Straub, *Die Geschichte der Bauingenieurkunst, op. cit.*, 160-161 (engl. : 2nd ed., Massachusetts, 1962, 116).

rings. It is this mechanism which may be called a scientific model in the present view.

I am going to put these four abstractions in more general context (table 1). The most important abstraction is formed by drawing (a/b). Yet between a building and its model, several other levels (1./2./3.) are being developed. From a building we arrive at a structure by taking account of the supporting parts such as arches, columns, slabs, foundations, and by omitting anything which does not carry such as tiles, plaster, windows and doors. From the structure we proceed to the structural model by drawing the elements which are considered necessary to show forces and deformations, stresses and strains. And from the structural model we abstract by describing those laws of physics which seem to be essential in a specific case. Primarily mechanical correlations are used and expressed geometrically or algebraically. At the beginnings of structural statics, *i.e.* in the 19th century, it was mostly done geometrically, whereas this days the numeric way is preferred thanks to digital calculation capacity.

Since the 18th century the idea of a structural model — illustrated by the mechanism in the example of the cupola — has been developed to a large variety[7]. The engineers have enlarged this theoretical level, besides traditional forms of buildings and structures, of course. On this level they have formed numerous models like beams, arches, frames, trussed girders, slabs, deep beams, shells and even models for free-shaped buildings (fig. 3). A particular instance is the trussed model, where the illustrations of building, structure, structural model and even of the calculation nearly look the same (fig. 4).

Meanwhile such structural models have become categories of design. From the engineering point of view, however, it is crucial to distinguish consistently between the different levels of abstraction. Every level should be open to individual development, otherwise the engineer's creativity would be heavily restrained. Yet such an admission may lead to open controversy between theorists and practitioners. As it is shown in history, it can even degenerate into polemic dispute at times[8].

THE SIGNIFICANCE OF TRADITIONAL MODELS

A second example shows the use and significance of traditional models. It is entirely different from the first one in so far as the scientific models and their level of abstraction are still missing. In 1755 the town-council of Schaffhausen invited tenders for the construction of a new bridge across the river Rhine. To secure this commission, the well-known master builder Johann Ulrich Grubenmann (1709-1783) presented the town-council with a wooden scale model of

---

7. H. Duddeck, " Entwicklung der Berechnungsmodelle des Bauingenieurs : Woher ? Wohin ? ", *Bautechnik*, 70 (1993), 640-649.

8. H. Straub, *Die Geschichte der Bauingenieurkunst, op. cit.*, 162.

the bridge. When he was ridiculed with the remark that such a bridge would surely not hold up, he jumped on the model with his whole weight and said : " If this model can carry my whole body, the real bridge will surely be strong enough to carry a few carriages ! " Grubenmann did secure the contract, because he succeeded in convincing the panel of negotiators[9].

Grubenmann and his people did indeed build the 18th century bridge. Although the spectacular securing of this commission is a mere anecdote, it is handed down by Swiss engineers to this day. It illustrates the fact that exceptional objects were built without abstract models. Mere traditional models like small-scale copies were sufficient to evaluate a project. Yet this anecdote shows how one can succeed economically with the help of models. Grubenmann's success did not primarily depend on the scientific value of his model, but on the notions of each individual involved.

These two examples are both well-known in civil engineering, but civil engineers are not explicitly mentioned. The first man to call himself a civil engineer was John Smeaton (1724-1792), an English instrument maker[10]. In 1771 he had founded the Society of Civil Engineers, which was followed by the Institution of Civil Engineers in 1818 and by similar associations in other countries later on. At the same time this new profession became closely bound up with the development of academies and schools for higher technical education.

A NEW MODEL OF THE BUILDING PROCESS MANAGEMENT

Designing by the use of scientific models is not the only characteristic of modern civil engineering. No less important is the new position adopted by the engineer when setting up technical facilities. These tendencies are similar to those in other branches within the building industry, and they show overlapping professional activities which are nevertheless autonomous being in the hand of specific professional associations. For example from England it is reported how the role of designing and surveying is strictly separated from trading and contracting. From this distinction the profession of the modern architect has developed[11]. The modern architect is no longer paid for an entire work, but he charges only the fee for his own service. A similar position is created by the engineers. Whereas the architects are concerned with buildings, civil engineers design structural facilities of several kinds " by the art of directing the great sources of power in nature for the use and convenience of man "[12]. The above mentioned public works are the principal parts.

9. J. Killer, *Die Werke der Baumeister Grubenmann*, 2. Aufl., Zürich, 1959, 21-31.

10. *The New Encyclopaedia Britannica*, vol. 18, 15th ed., Chicago, 1994, 416.

11. H. Colvin, *A Biographical Dictionary of British Architects 1600-1840*, 3rd ed., New Haven, London, 1995.

12. Royal Charter of " The Institution of Civil Engineers ", incorporated June, 3, 1828, *List of Members*, London, 1867.

In fig. 3 there is a sketch of this new management structure. There is a need or a request to be satisfied by a work and an idea of how to build this work. At the beginning none of these three elements is known exactly. Client, engineer and contractor negotiate. In this structure, an independent engineer is responsible for the design of a plant and he writes a technical report. He explains his ideas to the client, and then publishes an illustrated, printed report. Afterwards he produces the detailed designs, the specifications and the cost estimate. During execution he does the supervision. The day-to-day direction may be the task of the so-called resident engineer, " a term introduced by Smeaton in 1768 "[13]. To evaluate his ideas the engineer gradually uses scientific methods and models. Thanks to his experience gained in this way, he may be commissioned as an expert of further projects.

This new management structure has remained unchanged until today and has become typical for building processes. It is to be added that — instead of a single person — today there are large groups of planners such as teams of engineers and architects. This structure is so characteristic that it has become a model too. Strictly speaking we have been allowed to call it a model only since the building process was made an object of theory too. Thanks to this model the role of civil engineers can be explained and — based on it — these explanations can be examined by a third party. Of course, this model is merely one among others. Yet it demonstrates another aspect of the term model and of its enlargement.

Let me sum up : in the 18th century scientists were confronted with building problems. With the help of their thoughts they enlarged the analysis of buildings by the addition of a new theoretical level. However, it was the engineers who developed the scientific models for their own purposes later on. So they turned them into a feature of their profession. Civil engineering is called an " applied science " and may be considered of secondary importance, but the development of its models shows it to be much more than a mere transfer of knowledge from science to engineering. Today there is a tendency to overestimate the importance of scientific models. As civil engineering cannot be dissociated from architecture, technology, science, economics and politics, it would indeed be much more appropriate to benefit from the rich and complex variety of models.

13. A.W. Skempton, *Civil Engineers and Engineering in Britain, 1600-1830*, Aldershot, Variorum Collected Studies Series : CS533, 1996, XVI.

FIGURES

1. Collection of models :
Physical Laboratory of the *Académie des Sciences* of Paris, 1711[14].

14. E.S. Ferguson, *Das innere Auge. Von der Kunst des Ingenieurs*, *op. cit.*, fig. 5.15.

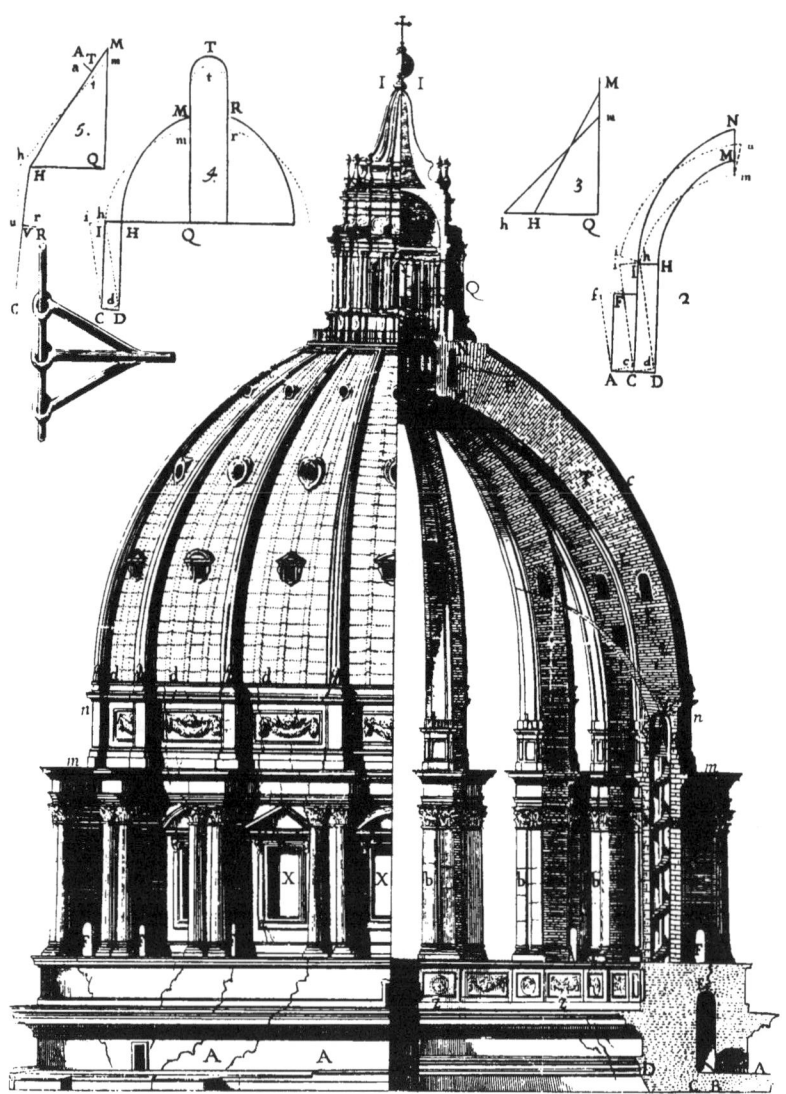

2. Cupola of S. Peter's Dome, Rome.
Plate from *Parere di tre mattematici sopra i danni che si sono trovati nella Cupola di S. Pietro sul fine dell'Anno 1742*[15].

15. H. Straub, *Die Geschichte der Bauingenieurkunst, op. cit.*, fig. 42.

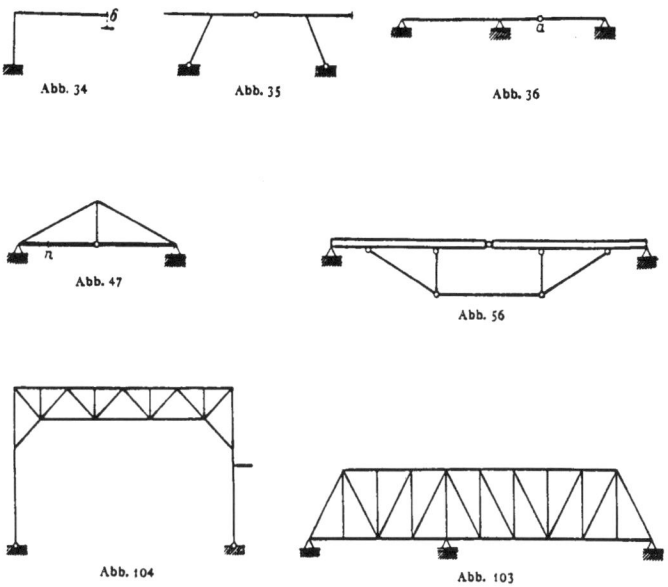

3. Structural models : 7 examples out of a large variety[16].

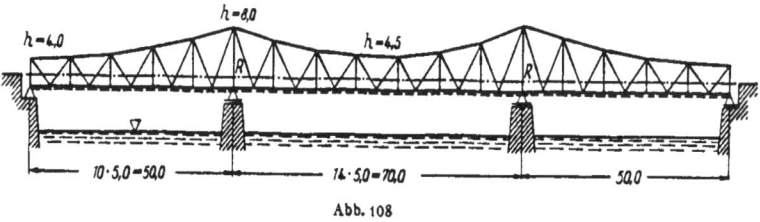

4. Trusses : high similarity of building, structure and statical model[17].

16. K. Hirschfeld, *Baustatik, Theorie und Beispiele*, 2. erw. Aufl., Berlin, 1965, 1025-1038.
17. *Idem,* 1031.

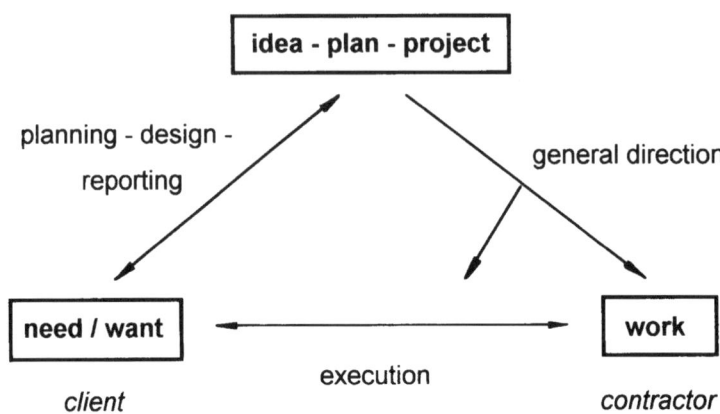

5. Management structure as a model.

TABLE 1.

MODELS AND THE LEVELS OF ABSTRACTION (A/B AND 1./2./3.)

Level	Object (a)	Model (b)
1	building - *bâtiment* - Gebäude	model - *modèle* - Modell / drawing - *dessin* - Zeichnung
2	structure - *structure porteuse* - Tragwerk	working drawing - *dessin d'exécution* - Konstruktionszeichnung
3	structural model - *système statique* - statisches Modell	static calculation - *calculation statique* - statische Berechnung

CHANGING IMAGES OF EDISON IN BIOGRAPHIES
AND THEIR SOCIAL CONTEXT

Shigeo SUGIYAMA

Thomas Alva Edison was born in 1847 and we are celebrating 150th anniversary of his birth this year. A number of biographies of him and articles on him have been published in these few years and some are to be published in the near future, in Japan as well as in the United States, to celebrate his achievements. Exhibitions displaying his great works are also being held in Japan.

The table below gives a non-exhaustive list of biographies of Edison published so far in Japanese.

YEAR OF PUBLICATION	AUTHOR AND TITLE
1908	T. Tanaka, *A chronicle of Edison's sayings and doings*
1929	K. Sawada, *A life of Edison*
1929	H. Makio, *Edison, the invention king*
1942	Y. Adachi, *Edison*
1945	— World War II ended —
1949	K. Sawada, *Edison*
1954	N. Sakikawa, *Edison*
1968	K. Nomura, *Edison*
1980	Y. Katagata, *Edison*
1981	S. Ohno, *Edison*
1981	N. Sakikawa, *Edison, a mischief and a genius of invention*
1994	T. Kimura, *Invention war : Edison vs. Bell*
1997	M. Yarnakawa, *An illustrative encyclopaedia of Edison*
1997	K. Hamada, *Edison, the genius : the forerunner of the 21st century*

Now, I will show in this paper how greatly the image of a scientist described in a biography written in a certain period of time is different from that described in another biography written in a different time. I will show it by taking the case of Edison as an example. I will also argue that various images of Edison that appear in different biographies written in different times were closely related to the social conditions and cultural atmosphere of the day when biographies were written.

It is not my concern here to ask to what extent an image of Edison represented in a biography differed from that of the real Edison. And I will pay attention primarily to biographies written by Japanese authors, because my concern here is how Japanese people thought of Edison. (Biographies of Edison translated from foreign books are, therefore, excluded in the table.)

To make my argument clearer and simpler, I will make those three books which were published before 1930 be represented by the book *Edison, the invention king* by Makio. A number of biographies published after World War II and before the 1980s can be represented by the book *Edison* by Sakikawa. As *Edison* by Adachi is the only biography of Edison published in Japan during World War II, I will take it into consideration. I will exclude those books which were published after 1980 from my consideration in this paper for the sake of simplicity. What I want to say can, nevertheless, be successfully communicated to the reader.

Now let us consider these three representative books (boldfaced in the table) one by one.

The Edison described in *Edison,* the invention king is typically a man of efforts. For example, the story goes like this. Edison looked day after day for the material suited for the filament of an electric bulb so that it could emit incandescent light for sufficiently long hours before burning out. He tried as many as a few thousand materials one after another. He worked hard for many years with less than four hours of sleep to find out the best material. After strenuous efforts, he finally hit upon a fiber of bamboo, which led him to the great success.

This image of Edison as a man of efforts impressed many Japanese with the following phrase which we are told Edison uttered in a certain occasion : " Genius consists of 99% of perspiration and 1% of inspiration ".

The Edison described in the book was also a man who created a new era of electricity when electric power took the place of steam power or water power in a variety of industries. In Japan, growing use of electricity in industries started in full-scale in the 1920s. That is, the 1920s marked the beginning of a new era of electricity in Japan.

The situation produced an effective use of the image of Edison. The image of Edison that he was the man who opened up the era of electricity was created mainly by those who ran electric businesses. They wanted to promote their business by the image of Edison. An example is the advertisement in an issue of October 1919 of a journal named *OHM* (The journal included a variety of popular articles concerning electrical science). This advertisement aimed to promote the selling of storage batteries which were so light and so small as to be fitted on a car to supply it with driving power.

Now, let us turn to the second book written by Adachi and published in 1942. In Adachi's book, Edison was depicted as a great entrepreneur as well as an inventor. Edison in this book is not simply the inventor of the electric bulb. It is stressed that he built up a total system of supplying electricity. We are told that Edison devised an electric generator and an efficient way of transmitting electricity, determined the supply voltage as 110 volts, and invented an electric bulb and a meter to measure the amount of electricity supplied. Electric bulbs constituted only a small part of the system he invented.

Moreover, it is stressed that Edison himself established a company and successfully carried on the business against gas suppliers. In this connection much attention is paid in the book to Edison's policy of patents. The book explains how Edison effectively defeated competitors by utilizing his patents as his weapons.

What is more, Adachi emphasizes that utilizing inventions in factories and promoting industrial development would eventually lead to the expansion of national strength. The opening sentence of the book is illustrative :

" We must admit that the United States, though it is a hostile country, has built up a great material civilization. I believe that we Japanese should investigate how and why the United States built up its material civilization and what it is based on. I therefore tried to examine in this book, taking the case of Edison as an example and from a rather technological point of view, what the condition was in which Edison could make great inventions ".

In those days Japan was deploying troops in China and the countries of Southeast Asia, and was at war with the United States. Therefore, inventions were greatly encouraged in Japan to strengthen national power and to gain eventually a victory in the war. The Royal Association of Inventions organized a series of activities to stimulate inventions in the 1940s. As one of these activities it published in 1941 a book entitled *Stories of Inventions*. In the first page of the book appeared a picture of Edison. This impressively exemplifies that Edison was, in these days among Japanese people, indeed a symbol of both invention and national strength based on it.

Lastly, I will discuss a book written by Sakikawa. World War II ended in 1945. Nine years later, the book was published. The image of Edison held by Japanese people changed drastically through the experience of defeat in World War II. The Edison described in the book is a peace-loving man and a great inventor who contributed not to strengthen national power but to promote human welfare through his many inventions.

It was described in the earlier books by Makio and Adachi that Edison played an important role in the Naval Advisory Committee during World War I and he himself invented various things. A kind of sonar to detect hostile submarines was often referred to as a representative and successful one. We are now told in the book by Sakikawa, however, that it was not Edison but Coolidge who invented it. What an astounding change it is ! We are also told that Edison accepted the directorship of the Advisory Committee with the hope of leading the war to an end as soon as possible to relieve people from miseries caused by the war.

It is also to be noted that references to Napoléon Bonaparte are totally absent in the book. Edison had been likened in many respects to Napoléon Bonaparte in the books published before and during World War II : Edison had as little hours of sleep as Napoléon ; Edison easily got angry like Napoléon, and Edison soon became to be as if nothing had happened, again like Napoléon ; Edison bore a strong resemblance in his appearance to Napoléon, and so on. Napoléon Bonaparte had been greatly admired by Japanese people for his military prowess till the end of World War II.

After World War II ended in 1945, Japan enjoyed a period of democratization. The new democratic Constitution of Japan was established, legal codes were largely changed to support gender equality and lessen the authority of Japan's traditional patriarchal family system, Land reforms were carried out, and the educational system was revamped. In schools, both the spirit of article 9 of the new Constitution which states that Japanese people forever renounce war and the importance of contributing to the promotion of human welfare were intensely taught. In this connection high regard was given to education in science and education in scientific reasoning. Thus we can see that the new image of Edison which appeared in the book by Sakikawa was created to meet the demand of the Japanese society of the day. Some biographies of Edison went so far as to neglect all the facts that Edison did commit researches for armaments. The cultural atmosphere of the day in Japan was the cause of giving no reference to Napoléon. With this modification some comic biographies of Edison as well as ordinary ones, both for school children, appeared.

The same as what I have described so far concerning Edison can be pointed out of Michael Faraday.

M. Faraday as described in a biography published in Japan in 1923[1] was typically a successful self-made man who went up from being in obscurity to a scientist of the first rank through persistent efforts. (It reminds us of Samuel Smile's *Self Help.*) The biography met the cultural atmosphere of the day when Japan was still an under-developed country and Japanese people were eager to emerge from poverty.

On the other hand, soon after Japan was defeated in World War II, M. Faraday was regarded as a great popularizer of science, who successfully communicated the essence of scientific thinking to common people. In those days the

1. K. Aichi, *Faraday, the great authority of electricity*, 1923.

Japanese government and people endeavored to drive away irrational or unscientific thought which was thought to be responsible for the militarism which had led Japan to the war, and to democratize Japanese society by spreading scientific thought among people. Science education was thus regarded as a means of democratizing Japan. A Japanese version of a Christmas Lecture was conducted in this connection in 1948, though once for all. The picture of Faraday as a man who led Christmas Lectures at the Royal Institution thus emerged in biographies published soon after 1945.

Biographies of scientists have described scientists' personal lives, their scientific careers and their achievements in science. However, descriptions differ appreciably from biography to biography depending on the time when they were written. It was not only because new facts of the scientists described were revealed, but also because, as the time went by, special attention was being focused, excessively in some cases, on another feature of the same scientist, the feature having been perceived, and yet regarded as not important before.

Biographies of scientists have been written and rewritten to meet the cultural atmosphere and social demands of the day : scientific biographies helped government and scientific community recruit young prospective scientists by inspiring their interest in science ; biographies promoted the modernization of Japanese society by depicting the scientists' hard struggle against irrational ideas ; they enhanced national dignity by praising the achievements of Japanese scientists, and so on. Biographies of scientists were in this way socially " constructed " and " utilized ".

Historians of science have worked primarily on what scientists told and wrote. Historians of science can and should pay much more attention to what is told and written of scientists. I believe that an analysis of a series of biographies of scientists written for common people would give us a new perspective on the history of science.

INDUSTRIAL ENGINEERING STUDIES IN SPAIN IN THE 19th CENTURY

José M. CANO PAVÓN

Industrial engineering studies were started effectively in Spain in 1850, with the promulgation of legislation that led to the systematic organization of curricula at different academic levels. Several industrial schools were founded that taught different disciplines and issued various degrees up to and including higher industrial engineer (granted by the Madrid school). The administrative structure of schools and studies was continually modified at an early stage, particularly between 1857 and 1860. From then to 1867, all industrial schools except that of Barcelona were closed down owing to economic difficulties and, especially, their small numbers of students. This situation was propitiated by the lack of professional outlets in both the public and the private sector, most Spanish regions were scarcely industrialized at the time.

In fact, the school of Barcelona was the sole higher industrial engineering educational institution in Spain in the last third of the 19th century. However, elementary industrial engineering schools, so-called " schools of arts and crafts " existed at several Spanish cities during that period. At the turning point of the century, the " Regenerationist " Movement that arose after the Spanish war with the United States and the loss of Spain's overseas colonies in 1898 revitalized higher engineering education with the foundation of the school of Bilbao in 1899 and that of Madrid in 1901.

This paper deals with general aspects of the development of industrial engineering in Spain since 1850. It discusses human and material resources available at the schools, the books their students used and the academic and professional fruits they bore.

INDUSTRIAL ENGINEERING EDUCATION PRIOR TO 1850

The foundation of industrial schools in Spain in the mid 18th century was not unexpected. Since long before, an interest had existed in furthering the teaching of the then called " industrial arts ". Some official and private initiatives had arisen to fill this gap in Spanish education in an attempt at training technicians with high or intermediate skills and increasing the technical knowledge of industrial workers and managers.

The analysis of changes in Spanish industrial education since the mid 18th century reveals three distinct periods or stages that coincided with the enlightened, pre-liberal and liberal historical periods. In each period, industrial education was organized in accordance with the conceptual ideas of the time.

In the enlightened period, which spanned from the mid 18th century to 1820, a number of scattered initiatives arose at different Spanish places to solve specific problems posed by the growing demands of the market. Although craftsmen associations provided major vocational training, the industry that grew outside them demanded new teaching institutions that could disseminate technical novelties among workers.

The most outstanding initiatives of this period stemmed from naval consulates and local market councils. The most active of these bodies was the Barcelona Market Council, which founded a School of Design in 1775, a School of Chemistry in 1805 and a School of Practical Machinery and Mechanics in 1808. Textile schools were started in other Spanish cities such as Valencia, Coruña and Burgos, and a school that taught naval, drawing and agricultural subjects was also created in Málaga during this period[1]. A government initiative instituted the Royal Cabinet for Machines, which operated from 1791 to 1824 and was set up by several technicians including Agustín de Bethencourt and Juan López Peñalver, who were previously trained in civil engineering at the *École des Ponts et Chaussées* of Paris on subsidies from the Spanish government[2]. The Cabinet subsequently served as an industrial museum.

The pre-liberal period of industrial education in Spain spanned from 1820 to 1845. The government undertook the task, with the help of enlightened bodies — particularly " economic societies ", which existed at many places in Spain. There was the widespread belief that, because of its special nature, this type of education should be provided at institutions other than the universities and that the French model of special schools should be adopted to this end. In 1824, the Art Conservatory of Madrid was founded after the *Conservatoire des Arts e Métiers* of Paris ; the Conservatory encompassed the Royal Cabinet for Machines, which was since then concerned with teaching as well. Teaching activities were expanded in the 1832-1833 academic year, when the Conservatory started to teach three different levels. Subjects included mathematics, mechanics, chemistry, draughtsmanship and machine building. Under the auspices of the Madrid conservatory, several chairs were created in cities such as Barcelona, Valencia and Málaga. Some (e.g. those of Barcelona and Valencia)

1. A. Escolano Benito, *Educación y economía en la España Ilustrada*, Madrid, 1988.
2. A. Rumeu de Armas, *El Real Gabinete de Máquinas del Buen Retiro*, Madrid, 1990.

led to the foundation of new conservatories[3]. As a rule, these institutions sought to train higher engineers. Simultaneously, the Spanish government sent some technicians out to France and Belgium for training as industrial engineers with a view to their serving as teachers in the projected new centres.

In the liberal period, which started in 1845 — or, effectively, in 1850 —, the Spanish government controlled industrial education in every respect. It established an organic, centralized model inspired by the French institutions and allocated significant economic resources to organizing the new schools and providing them with teachers. Those who were to teach the core subjects (mathematics, geometry, physics, chemistry) came from the university, where scientific disciplines had been taught since 1845 ; those who were to teach technological subjects, were engineers trained abroad. The earliest attempts at organizing industrial curricula were made in 1845 and 1850 in Gijón and Vergara, respectively ; however, it was not until pertinent organic laws were issued in 1850 that the new scheme, described in detail in the following section, was formally established.

DEVELOPMENT OF INDUSTRIAL EDUCATION SINCE 1850

In September 1859, the so-called " decree of Minister Seijas " was issued. The decree placed industrial education in a centralized framework and established three different study levels, namely : (a) elementary, which was taught at some secondary education institutes ; (b) intermediate, which was initially taught at the new industrial schools of Barcelona, Sevilla and Vergara, and subsequently also at Valencia ; and (c) higher, with a mechanical and a chemical speciality, both of which were taught at the Royal Industrial Institute of Madrid, built on the site of the former Art Conservatory. The intermediate schools also taught the elementary curriculum and the higher schools taught both lower levels. In addition to its teaching activities, the Royal Industrial Institute granted industrial privileges, issued patents and sustained an industrial museum. This educational scheme provided seven different degrees that ranged from skilled worker to higher industrial engineer. The studies were all free since Spain was in a pressing need for skilled engineers and technicians — most applicable jobs in major national industries and the incipient railway lines were being held by foreigners[4].

The industrial schools thus created started their activities between 1851 and 1852 ; their consolidation was a slow process[5]. During these years, the Royal

---

3. A. Rumeu de Armas, *Ciencia y tecnología en la España Ilustrada. La Escuela de Caminos y Canales*, Madrid, 1980.

4. J.M. Cano Pavón, *La Escuela Industrial Sevillana (1850-1866). Historia de una experiencia frustrada*, Sevilla, 1996.

5. G. Lusa, " La creación de la Escuela Industrial de Barcelona (1851) ", *Quaderns d'Historia de l'Enginyeria*, 1 (1996), 1-51.

Industrial Institute also operated a school for accelerated training of teachers for the new institutions. Also, some teachers were sent abroad to update their knowledge[6].

In May 1855, the first reform of industrial education in Spain was undertaken under the decree of Minister Luxán. It was aimed at improving and simplifying the previous structure of industrial schools by reducing the number of degrees awarded and establishing regulations on teaching activities, examinations and teaching staff provision. A new, intermediate school was founded in Gijón in 1856[7].

However, the 1855 reform was short-lived. Two years later, in 1857, a new regulation under the name " Law of Public Instruction ", by Minister Claudio Moyano, introduced a radical change in industrial teaching, which was brought closer to the university model. Elementary industrial teaching was since then entrusted to institutes, where it coexisted with secondary education curricula. Intermediate industrial studies were combined with higher studies to establish a unified higher industrial curriculum that was taught not only at the Royal Industrial Institute but also at the industrial schools of Barcelona, Sevilla, Vergara, Valencia and Gijón, which were appointed higher schools. Studies ceased to be gratuitous, which hindered access of the working class. In addition, the funding of all industrial schools except the Royal Industrial Institute was shared by the government and the provincial and local councils. This introduced some economic instability that hampered the subsequent functioning of the schools. Later legal regulations (1858 and 1860) altered the industrial teaching curricula[8].

These reforms introduced major changes in the general scheme of industrial engineering education in Spain. In 1860, the Vergara and Gijón schools were closed down because the provincial and local authorities refused to contribute to their support. Also, the few professional outlets for industrial engineers in Spain in both the private and the public sector — a result of the country being scarcely industrialized in the former case and the government granting industrial engineers little say in industrial management and rare access to public offices in the latter —, led to an abrupt drop in the number of students (e.g. there were only 30-35 students at Seville and 15-20 at Valencia in 1864)[9]. Such short students roles, in addition to the relative high expenses incurred by the students, pressure from other engineering groups and the economic crisis of

6. J.M. Cano Pavón, " El Real Instituto Industrial de Madrid (1850-1867) : medios humanos y materiales ", *Llull*, in press.

7. J.M. Cano Pavón, " La Escuela Especial (1845-1855) y de Industria (1856-1860) de Gijón ", *Llull*, in press.

8. J.M. Cano Pavón, " La enseñanza de la ingeniería industrial en España entre 1850 y 1868. La Escuela Industrial de Sevilla ", *Llull*, 19 (1996), 27-50.

9. J.M. Cano Pavón, " La Escuela Industrial de Valencia (1852-1865) ", *Llull*, 20 (1997), 117-142.

the time, lead the government to gradually close the schools down (Valencia in 1865, Seville in 1866 and the Royal Industrial Institute of Madrid in 1867). Only the Industrial School of Barcelona survived, thanks to the support of Catalonian bodies and firms. The school of Barcelona was thus the sole industrial engineering school that operated in Spain between 1867 and 1899[10]. It attracted a number of teachers from the extinguished schools and reached a fairly high teaching level.

The exclusive role of the Barcelona school came to an end in 1899, when the Industrial Engineering School of Bilbao was founded after long negotiations. Two years later, the new school of Madrid was started. These two new schools were created soon after the 1898 war against the United Stated, which led to the loss of Spain's overseas colonies (Cuba, Puerto Rico and the Philippines)[11]. This historic disaster propitiated the emergence of a new social and political current known as " regenerationism " that sought the progress of Spain via the modernization of its economic structures and the promotion of science and technology as the vehicles for solving the country's age-old problems.

HUMAN AND MATERIAL RESOURCES IN THE SPANISH INDUSTRIAL SCHOOLS OF THE 19th CENTURY

The teachers of Spanish industrial schools between 1850 and 1867 were partly industrial engineers trained in France or Belgium and partly graduates from the Royal Industrial Institute of Madrid. These engineers taught specialized technological subjects. Other teachers, including architects, pharmacists and mathematicians of different origins, were in charge of core subjects — particularly those taught in the first few years.

Staff composition varied from school to school. Thus, the staff of the Royal Industrial Institute of Madrid, which taught all three levels throughout its existence, consisted of 19 teachers and 7 assistants for practical lessons. The subjects taught in the two specialities (mechanics and chemistry) were algebra, trigonometry, descriptive geometry, general physics, general chemistry, draughtsmanship and industrial design, industrial building, applied inorganic chemistry, applied organic chemistry, chemical analysis, dyeing and ceramic arts, machine building, mechanical technology and arts, political economy, industrial law, French and English. The teaching staffs of the industrial schools of Seville, Vergara and Valencia were composed of 7 or 8 members, and that of Barcelona of 10 ; all four schools had 2-4 assistants for practical lessons.

10. R. Garrabou, *Enginyers industrials, modertnizació economica i burgesia a Catalunya*, Barcelona, 1982.

11. J.M. Alonso Viguera, *La ingeniería industrial española en el siglo XIX*, Madrid, 1961.

The school of Gijón, which started teaching the intermediate level in 1856, had fewer teachers ; also, its projected chairs were never occupied in full.

The staffs of the Barcelona, Valencia and Seville schools were slightly expanded when they were granted the right to teach the higher level (in 1859-1860). Some of the teachers at Valencia moved to Seville as the former school was closed down in 1865. Also, when the Seville and Madrid schools ceased to operate, some of their teachers were enrolled by the Barcelona school and some joined the science faculties of Madrid, Barcelona, Seville and Valencia.

The industrial teachers of the time included some prestigious figures. Thus, Ramón Manjarrés, who held the chemistry chair at Seville from 1860 to 1866 and the Barcelona chair between 1866 and 1891, was an expert in agricultural chemistry[12] and one of the first to introduce the industrial use of electricity in Spain. Magín Bonet Bonfill, professor of chemical analysis in Madrid and a disciple of Dumas, Fresenius, Bunsen and Berzelius, was primarily concerned with oenology. One other prominent figure was Francisco P. Rojas, initially professor at Valencia (1855-1865) and then at Barcelona, who worked on heat transfer processes and electric machines[13]. It is also worth mentioning Cipriano S. Montesinos, professor at the Royal Industrial Institute of Madrid between 1851 and 1854, who was a specialist in machine building, about which it published a comprehensive treatise[14] ; Montesinos was a member of the international committee that supervised the project of the Suez canal.

Prominent among Madrid teachers were Isaac Villanueva, who was the author of an extensive course of industrial drawing, and Professors Constantino Sáez Montoya and Luis M. Utor Suárez, who worked on the analysis of waters and on various agricultural chemistry topics. Finally, due mention should be made of Luis Justo Villanueva, professor at Gijón between 1856 and 1869, and later at Barcelona, who became an expert in the recovery of waste waters.

As regards material resources, the situation was markedly different among schools. Most of them — those of Vergara and Gijón excepted — were established in former convents that had been confiscated by the civil authorities. They obviously necessitated reformation works that were highly costly in some cases (e.g. Seville). As a rule, the schools possessed acceptable physical and chemical laboratories, and also well-equipped drawing rooms furnished with extensive collections of plates. Machine workshops were relatively poorly equipped ; the lack of space and budgetary restrictions prevented the schools from having self-contained workshops in the 1850-1867 period. The workshops usually included machine models ; however, life-size machines were

12. J.M. Alonso Viguera, *La ingeniería industrial española en el siglo XIX*, Madrid, 1961.

13. J.M. Cano Pavón, *La ciencia en Sevilla (siglos XVI-XX)*, Sevilla, 1993.

14. F. de P. Rojas, *Estudio matemático de las máquinas magnetoeléctricas y dinamoeléctricas*, Madrid, 1887.

scanty and so were small machines and lathes for mechanical work, as well as the steam engines required to power them.

The Royal Industrial Institute of Madrid possessed an industrial museum that held extensive collections of industrial products and raw materials, as well as many models and plans. Library collections were largely acceptable ; most schools possessed 300-500 volumes of general and specialized books that were predominantly in French. The library of the Gijón school contained more than 6.000 volumes, most of which, however, were humanistic and legal works donated by the enlightened politician Gaspar Melchor de Jovellanos (1744-1811).

INDUSTRIAL ENGINEERING TEXTBOOKS

Like students at university faculties, those at industrial schools used the textbooks periodically imposed by the government for use all over Spain. The books most frequently used for the teaching of the different subjects over the 1850-1870 period were as follows :

-        Industrial physics : *Applied Physics Treatise*, by Peclet[15] ; and *Electricity Treatise*, by Fernández de Castro[16].

-        Industrial mechanics : *The Engineer's Handbook*, by Nicolás Valdés[17] ; and *An Industrial Mechanics Course*, by J.V. Poncelet[18]

-        Stereotomy : *The Engineer's Handbook*, by Nicolás Valdés ; *Stereotomy Treatise*, by Le Roy[19] ; and *Stereotomy Treatise*, by Adhemar[20].

-        Industrial building : *Industrial Building Treatise*, by Demannett[21] ; and *The Engineer's Handbook*, by Nicolás Valdés.

-        Steam engines : *The Engineer's Handbook*, by Nicolás Valdés ; *Steam Engine Treatise*, by Tredgold[22] ; and *Theory of Steam Engines*, by Guionneau de Pambour[23].

-        Chemical analysis : *Chemical Analysis Treatise*, by Rose[24] ; and *Chemical Analysis*, by Fresenius[25].

15. C.S. Montesino, *Resumen del curso de construcción de máquinas*, Madrid, 1854.

16. M. Fernández de Castro, *La electricidad y los caminos de hierro*, Madrid, 1857.

17. N. Valdés, *Manual del Ingeniero*, Paris, 1859.

18. J.V. Poncelet, *Traité de mécanique industrielle*, Liège, 1845.

19. C.F.A. LeRoy, *Traité de Stéréotomie*, Paris, 1844.

20. J. Adhemar, *Traité de géométrie descriptive*, Paris, 1946.

21. A. Demanet, *Cours de constructions*, Bruxelles, 1847.

22. T. Tredgold, *Traité des machines à vapeur et de leurs applications*, Bruxelles, 1838.

23. F.M. Guionneau de Pambour, *Théorie des machines à vapeur*, Paris, 1844.

24. H. Rose, *Traité pratique d'analyse*, Paris, 1843.

25. R. Fresenius, *Précis d'analyse chimique qualitative*, Paris, 1847.

-     Applied inorganic chemistry ; *Applied Chemistry Treatise*, by Payen[26] ; and *Applied Chemistry Treatise*, by Dumas[27].
-     Dyeing and ceramic arts : *Ceramic Arts Treatise*, by Salvetat[28] ; and *Chemistry as Applied to Dyeing*, by Persoz[29].
-     Political economy and industrial law : *Political Economy*, by Benigno Carballo[30].
-     Draughtsmanship and industrial design : *Industrial Design Treatise*, by Isaac Villanueva[31] ; and *Elements of Draughtsmanship*, by J.B. Peyronnet[32].

Most of these books were original versions in French, the reading of which demanded command of this language, which was thus taught by the schools themselves.

EXAMINATIONS AND END-OF-STUDIES PROJECTS

Industrial students were evaluated by examinations that differed in nature from those taken by university students. Thus, they had half-year examinations in February and whole-year examinations in June. Examinations were usually in the form of written papers ; by contrast, those at universities were mostly oral. As a rule, evaluations were not too strict as usually more than 50% of the candidates passed their exams — with marked differences among subjects, however.

In order to obtain their industrial engineer degrees (mechanics or chemistry branch), students had to pass a final examination before a teachers' board and then make the project for an industrial facility. They were allowed two months for this purpose, at the end of which they were to hand in a report and pertinent sketches and plans. Candidates were to design mid-level facilities that could be set up in the region where the industrial school that would issue their degrees was located.

Students at the school of Barcelona, which was an industrialized area, had easier access to industrial facilities ; as a result, many Barcelona graduates argued that they were better qualified than colleagues from other schools, which were located in predominantly agricultural regions (e.g. Seville, Valencia) or in the Madrid of the 19th century, which was an administrative and service city with little industrial weight.

26. A. Payen, *Précis de chimie industrielle*, Paris, 1859.
27. J.B.A. Dumas, *Traité de chimie appliquée aux arts*, Liège, 1847.
28. M. Salvetat, *Le cours de céramique*, Paris, 1857.
29. J. Persoz, *Traité théorique et pratique de l'impression des tissus*, Paris, 1846.
30. B. Carballo, *Curso de economía política*, Madrid, 1855.
31. I. Villanueva, *Dibujo geométrico aplicado a las artes*, Madrid, 1835.
32. J.B. Peyronnet, *Elementos de dibujo lineal*, Madrid, 1835.

TEACHING AND PROFESSIONAL BALANCE OF INDUSTRIAL SCHOOLS

The body of Spanish industrial schools that operated between 1850 and 1867 produced about 270 graduates, of which 165 obtained their degrees from Madrid. As noted above, the sole industrial school existing in Spain between 1867 and 1899 was that of Barcelona, which produced 744 engineers (528 mechanical and 218 chemical specialists) over that period. Overall, engineering graduates from Spanish industrial schools in the second half of the 19th century amounted to about one thousand.

Although there are no reliable data in this respect, many of such engineers were enrolled as teachers at universities, arts and crafts schools, and institutes. Another significant fraction was engaged by telegraphic services and gas suppliers. The remainder probably had industrial jobs, particularly in the iron and steel industry — then at an early stage of development —, or engaged themselves in the booming railway industry.

CONCLUDING REMARKS

Overall, industrial engineering education in Spain followed a winding path in the second half of the 19th century ; previously — from the late 18th century — industrial teaching had focused preferentially on workers and craftsmen in the most economically active Spanish regions.

Notwithstanding its excessively centralist and pyramidal nature, the teaching scheme started in 1850 and refined in 1855 had the advantage that it established three different study levels, viz. elementary, intermediate and higher, aimed at training skilled workers, technicians (*peritos*) and industrial engineers, respectively. The law of 1857 completely altered the initial scheme, which could hardly be applied during its years of enforcement. Elementary industrial studies were confined to secondary education institutes — which lacked the proper means and environment required for their new task — and the intermediate level was suppressed — incorporated into the former higher studies. These, together with the shortage or professional outlets, gave rise to an abrupt decrease in the number of students that ultimately led the government to close down most of the schools.

Industrial schools were expensive to run in most cases. In fact, the total investment between 1850 and 1867 was estimated to be 25 million reals of the time (1 real = 0.25 french franc in those years), which was quite substantial relative to university costs. On a relative scale, the annual tuition cost of a higher industrial education student was 7.000 reals, compared to 1500-1800 reals for a student at the Civil Engineering School of Madrid, and less than 1.000 for one at a university faculty.

The development of industrial schools in Spain bears witness to the failure of an educational policy that proved unable to develop a suitable model for the prevailing social and economic situation of the country and in the end ruined the efforts made between 1850 and 1857 to create a school network in response to the growing demand of Spain's industrialization process for technicians of different skill levels.

THE CONTRIBUTION OF THE CIVIL AND MILITARY ENGINEERS
IN THE DEVELOPMENT OF THE MACHINES' MECHANICS
IN THE 19th CENTURY IN RUSSIA

Margarita M. VORONINA

At first, I will try to find the influence of 18th century's mechanics on the 19th century's mechanics in Russia. In fact this influence was not very large. We can see the development of practical mechanics but theoretical mechanics were not born yet. Only some célèbre scientists are known but they did not leave their own scientific schools. We know that Leonard Euler (1707-1783) had not brought up any well known scientist. Mikhail Lomonosov (1711-1765) had not established his school, although he made a program of studies on mathematics and mechanics for military high-schools. This program was enough full and nice and had some influence on the forming of the some courses. To the end of 18th century in Russia there was no leaders in mechanics (however in France there was a lot of them). Only several men were carrying it. They were S.E. Guriev (1763-1813), V.I. Viskovatov (1780-1812), D.S. Chizhov (1785-1852). So we had a rupture between practical and theoretical mechanic. In these conditions Russia — the enormous feudal and serfdom state — came up to the beginning of the 19th century.

We characterise the life in Russia in the first 30 years of 19th century as the development of industrial capitalism, increasing of the number of factories, learning the ideas of Napoleon's France. As the increasing of the industrial production and political and economic power of state was impossible without any progress in many different branches of science, the government took for itself the role of patron of science and education. We should note two other moments : the industrial revolution in Russia as in the others countries was made by the introduction of the steam engines in the factories, manufactures and city traffic. However there was a special problem in our country — the problem of large distances. That is why it was necessary to build the whole net of water and ground means of communication. As a result, Russia met the same problem as France at the end of the 18th century : the necessity of self scientific staff, self scientific and technical intelligence. In France they found a solution by creating the *École Polytechnique* and in Russia the same role was played by the Institute of Corps of engineers of means of communication (now — St. Petersburg State University of Means of Communication). It was founded in 1809, and for many years it was the science and technical centre in Russia, particularly the research centre on applied mechanics, the main meaning of which during the first half of the 19th century was mechanics of machines.

From its beginning the Institute was considered as a higher technical educational institution. Further it became the centre of research on mechanics. That can be explained not only by its purposes, but also by the fact that the main part of the works on mechanics was made in it by its professors or graduates. In particular the applied mechanics and building mechanics were born in the Institute and formed there as new sciences.

The first director of the Institute of Corps of Means of Communication was Spaniard Augustine Betancourt (1758-1824), the famous Professor of mechanics and builder[1]. I think it will be interesting to know that in Spain in 1996 and in Russia in 1997 was organised the impressive exhibition devoted to Augustine Betancourt.

When Betancourt was the head of the Institute, he invited for teaching work the well-known Russian Professors of mechanics S.E. Guriev, V.I. Viskovatov, D.S. Chizhov, the military engineer A.I. Maiorov (1780-1848) (unfortunately the two first scientists worked at the Institute for a little time), and also French engineers — P. Bazaine (1786-1838)[2], Ch. Potier (1785-1855)[3], A. Fabre (1782-1844), M. Destrem (1788-1855), and a little later G. Lame (1795-1870)[4] — and B.P.E. Clapeyron (1799-1864). Maiorov, Betancourt, Chizov, Lame and Clapeyron were first in Russia to study the questions of applied mechanics, in particular the question of mechanics of machines.

In 1812 Maiorov began to read the course of lectures on mechanics at the Institute, and he included some elements of the mechanics of machines in this course. In 1817 Basen published a book about the motion of steam boats also with some elements of calculus of machines. Five years later the first book in Russian of machines mechanics by Chizhov was published (entitled *Notes on the application of base of mechanics to the calculus of working of some machines, the most usable*). This book was similarly to first works of J.-V. Poncelet by its maintenance and its methods. From 1820 the course of mechanics at the institute was read by Clapeyron. He added the sections about calculus of working machines, about steam and water machines and about wind engines.

1. A.N. Bogoljubov, *Avgustin Avgustinovich Betankur*, Moskva, 1969 (in Russian).
2. D.Ju. Guzevich, I.D. Guzevich, *Petr Petrovich Bazen*, St.-Petersburg, 1995 (in Russian).
3. B.F. Tarasov, *Jakov Aleksandrovich Sevastjanov*, St.-Petersburg, 1995 (in Russian).
4. M.M. Voronina, *Gabriel Lame*, Leningrad, 1987 (in Russian).

In 1823 the " teaching about engines and reception of working forces " was made as a special subject, named " applied mechanics ". So, firstly in Russia, applied mechanics was studied as a separated subject, on which from the 1824 the students passed the exams each year. In 1824 the lectures of Clapeyron were duplicated. Now it is possible to see these lectures in the library of the University of Means of Communications in Petersburg.

Thus, the first textbooks for the new course of applied mechanics were the works of Maiorov, Bazaine, Chizhov, Clapeyron and some articles of Lame.

Further development of applied mechanics as a science was connected at first with the names of M.V. Ostrogradsky (1801-1861/62) and P.P. Melnikov (1804-1880). The well-known Professor of mathematics and mechanics M.V. Ostrogradsky entered at the Higher Technical School in 1830. Before this time he passed some years in Paris, where he worked with the eminent scientists : A. Cauchy (1789-1857), S. Poisson (1781-1840), J. Fourier (1768-1830).

From the 1830s in Russia there was an intensive development of applied sciences, that was closely connected with the large development of technics and rail roads construction. Ostrogradsky had the lectures in five educational institutes of Petersburg, but he especially marked the Institute of means of communication, where he worked from 1830 till the end of his life. He kept the post of Main Educational Observatory of mathematical sciences in St.-Petersburg and he exerted influence on the education process in mathematical department in Moscow University and in Moscow Handicraft Institute (later — Moscow higher technical institute named by Baumann). That means that the influence of Ostrogradsky on the education of mathematical subjects was spread to the whole Russia. We have to note that at that time not only mathematics but also mechanics and applied mechanics were named as the mathematical subjects.

Ostrogradsky especially paid attention to the connection of mathematical sciences with the practice, with engineering art in his concrete works. This position was taken by his students, for example by P.P. Melnikov, S.V. Kerbedz (1810-1899)[5], I.A. Vyshegradsky (1832-1895), N.P. Petrov (1836-1920) and others, in particular in the field of applied mechanics.

We will look at his students and collaborators in the Institute of means of communication. P.P. Melnikov[6] was graduated from this Institute and he was left there to read the lectures on applied mechanics. He spread the course of Clapeyron and he connected this course with the course of analytical mechanics, which was read by Ostrogradsky. Melnikov almost created his own course

5. M.I. Voronin, M.M. Voronina, *Stanislav Valerianovich Kerbedz*, Leningrad, 1982 (in Russian).

6. M.I. Voronin, M.M. Voronina, *Pavel Petrovich Melnikov*, Leningrad, 1977 (in Russian). See also : M.I. Voronin, M.M. Voronina, *Pavel Melnikov and the Creation of a Railway System in Russia, 1840-1880,* Danville (PA), USA, 1995.

and he made the revolution in this teaching not only in Petersburg but in the whole country.

In his first book *About railroads* (1835) he generalised the last achievements of English and French scientists and he added them to his own investigations. Melnikov gives a short description of steamengines and he proved that moving steam engines are the most economical engines for the rail transport. This book is particularly interesting by the fact that it was written in the time of strong discussions about the possibilities of railroads construction in Russia.

In 1837 Melnikov duplicated his lectures of Applied Mechanics where he wrote the main propositions about engines and machines. In these lectures at the first time in Russian science literature Melnikov introduced the notion of work as a multiplication of force and distance (before it was named moment or dynamical moment), he gave an explanation of the parallelogram of J. Watt (1736-1819) and he talked about the works of G. Prony (1755-1839) and T. Tredgold (1788-1829) in this field. Melnikov was the first among Russian scientists who was interested by calculus of straight-directional mechanism of Watt and he introduced this part in his course of applied mechanics.

The Institute of Means of Communications played an important role not only in the creation of applied mechanics as a science but also in the distribution of knowledge in this field. In all technical institutes of Petersburg and even in Petersburg State University the courses of applied mechanics were given by the Engineers of Means of Communications, namely the students and collaborations of Melnikov. They were N.F. Yastrzhembsky (1808-1872), F.I. Sulima, P.I. Sobko (1819-1870), F.I. Enrold, Kerbedz. The lectures of Kerbedz were listened by G.E. Pauker (1822-1889) and Petrov — the well-known Russian Professors of mechanics.

In Russian universities applied mechanics received an official status only with the order of 1863, in which a special chair of applied mechanics was provided but education of some chapters of applied mechanics appeared in the universities only at the end of the thirties.

In the 1840s some new works on mechanics of machines were printed. The first of which were the works of Yastrzhembsky, N.N. Bozheryanov (1811-1876), A.S. Ershov (1818-1867). The engineer N. Yastrzhembsky was the first who introduced a course of machines construction as a whole subject in Technological Institute. He wrote some textbooks and drawings for this course, and published several books on the question of applied mechanics and the mechanics of machines. In the first time at Russian science literature he marked the electricity as a potential engine.

At the end of the 40s the most important part of applied mechanics was the theory of machines, and in particular the theory of steam engines. The investigations in this field were made also by Bozheryanov and A.G. Dobronravov. Bozheryanov, the student of Ostrogradsky in the Naval Corps, was a lecture on applied mechanics. He had some work in this field, in particular *The theory of steam engines* (1849) which was awarded by the Russian Academy of Sciences. The engineer of means of communications Dobronravov also published some works on the theory of steam engines. The most famous are *General Theory of Steam Machines* and *Theory of Steam-engines*.

The course about the transmission and transformation of motion started in the 1860s in the Institute of Means of Communication. That means that in the beginning of the 60s some attempts of cine-statics calculus of mechanisms were made. These years Ershov, graduated from the Moscow University, published in Moscow a part of his lectures on applied mechanics, namely *The Foundations of Cinematic or Theory Elementary about Moving in General and Moving of Mechanisms and Machines in Particular* (1854). It was the first book in the Russian language devoted to the problems of technical cinematic.

The works in the field of machine mechanics were continuing by Vyshnegradsky — a student of Ostrogradsky. He gave the lectures on applied mechanics in some institutes of St.-Petersburg and also he gave the lectures on Constructing of machines, theoretical engineering, theory of elasticity and thermodynamics. His main branch of investigations was the theory of regulation of motion of machines. In the end of 1870s he founded the beginning of the engineering theory of regulations. His ideas were developed later in the works of Koturnitsky, A.I. Sidorov (1866-1931), and later N.E. Zhukovsky (1847-1921), V.L. Kirpichev (1845-1913) and K.E. Rerih (1878-1934).

Vyshnegradsky made a valuable contribution to the development of practical mechanical engineering in our country (rearmament, reconstruction of plants, works on construction, projecting). He also took part in the running of rail roads and industry. He was not only a scientist and teacher but also an organiser. He worked a lot on the problems of technical education in Russia. In 1884 he made a project of development of professional education in our country, in which he provided not only preparation of engineers but also of technicians, masters and workers.

In 1871 he founded the group of scientists who worked in the field of applied mechanics. This group was named " Pentagonal Society " and it was composed of five members who met every week. In this group were the well-known Professor of mechanics Petrov, V.L. Kirpichev (1845-1913), A.P. Borodin (1848-1898) and Koturnitsky who had great influence on the development of applied mechanics. For example Petrov studied in St.-Petersburg and he was the student of Ostrogradsky and Vyshegradsky. His works were devoted to the different fields of applied mechanics — theory of tooth gearing, speed of trains, theory of traction. But the main field of his investigations was the hydrodynamic theory of greasing, in fact he founded this theory. In 1883 he published his well-known work *Traction in machines and the influence of grease on it.*

As a result, by summing all of this, we can see that in the first 2/3 of the 19th century in Russia was founded the base of investigation of machines mechanics. The main role in this was played by the civil and military engineers.

Paul Voïnarovski et l'Institut Montefiore

Véra Sévérinova et Constantin Sévérinov

Paul Voïnarovski (1866-1913)

C'est par un jour de février que l'on a apposé une plaque commémorative à Paul Voïnarovski, sur le mur du bâtiment principal de l'Université d'électrotechnique à Saint-Pétersbourg. A l'université, les cours consacrés au jubilé étaient organisés par les ingénieurs et scientifiques de Saint-Pétersbourg. Ce fait est remarquable, puisque le nom de Voïnarovski n'avait pas été mentionné depuis longtemps. Son activité polyvalente en électrotechnique était largement appréciée par ses contemporains mais oubliée par la génération suivante. Son nom n'est cité que dans quelques travaux universitaires de sciences naturelles, de technique, dans des ouvrages du Musée Central de la Communication Popov et dans un article de la Grande Encyclopédie Soviétique. En 1994 Catherine Popova-Kiandskaia, la petite-fille de l'inventeur de la radio, Alexandre Popov, publie un article qui marque le début de la renaissance de la mémoire de Voïnarovski.

Paul Voïnarovski occupe une place prédominante dans l'histoire du développement de toutes les sphères électrotechniques russes. Dans le dernier quart du XIXe siècle et au début du XXe, l'électrotechnique a subi en Russie comme dans le monde entier, un essor considérable. Les domaines de la science, de la pratique, de la pédagogie, de la formation des cadres qualifiés (ingénieurs et électrotechniciens) sont également touchés par ces changements révolutionnaires. Le travail de Paul Voïnarovski se retrouve dans tous les domaines qui restent marqués par son passage.

Paul Voïnarovski est né le 15 février 1866 dans le port de Sébastopol dans une famille de professeurs russes de littérature. Il passe sa jeunesse dans un autre port, celui de Marseille où il a terminé le lycée et la faculté de physique et de mathématiques de l'Académie d'Aix-en-Provence. Du sud de la France, le jeune licencié déménage sur les côtes de la mer Baltique à Saint-Pétersbourg et commence ses études dans le tout récent collège de la Poste et des Télégraphes (ce collège a été rebaptisé en 1891 Institut d'électrotechnique - premier institut supérieur de l'électrotechnique en Russie). A la fin du collège en 1890, il commence ses travaux pratiques à Moscou dans les bureaux de télégraphes. A cette époque, il mène déjà une activité civilisatrice. Il donne des conférences sur l'électricité et la construction du réseau électrique, prend part à l'organisation de l'exposition électrotechnique à Moscou et collabore au journal *Nouvelles Techniques* (industrielles). En même temps il soutient brillamment sa thèse et reçoit en 1894 le titre d'ingénieur en télégraphie (le sujet de sa thèse est l'illumination électrique des établissements d'État, des théâtres, des rues principales, des grandes artères ainsi que des maisons privées dans le centre de Moscou).

Paul Voïnarovski comprend le désir des ingénieurs russes de développer l'industrie du pays : " L'application de l'énergie électrique à l'industrie, écrit-il, a atteint des dimensions immenses. Mais est-on prêt à satisfaire ce besoin qui augmente ? " En répondant négativement à sa question, il décide de se consacrer à la pédagogie.

Le directeur de l'Institut d'électrotechnique Nicolaï Pissarevski a communiqué au chef de l'Office principal de la poste et des télégraphes : " En considérant les grands talents de Voïnarovski et en particulier ceux de sa pédagogie, je trouve non seulement désirable mais surtout utile pour notre travail de profiter de la proposition de Voïnarovski qui a prouvé ses compétences de professeur et son savoir-faire ".

Néanmoins, Paul Voïnarovski dans son activité pédagogique éprouve le manque de matériel scientifique et de méthodologie pédagogique dans l'électrotechnique formée, mais encore jeune. Le besoin d'acquérir une expérience la plus large possible en ce temps-là en Europe et en Amérique était vital. Il a choisi l'Université de Liège pour se perfectionner dans la formation des cadres russes dans le domaine de l'électrotechnique. Peu de temps avant, dans cette ville, à la même université, avec l'aide financière de Giorgio Montefiore Levy, sénateur italien, et sous la direction du célèbre électrotechnicien Éric Gérard, a été fondé l'Institut d'Électrotechnique Montefiore. En automne 1894 débute son séjour dans cet institut et sa collaboration avec le professeur Éric Gérard, période courte d'un an mais significative dans la vie de Paul Voïnarovski. Heureusement les lettres écrites à cette époque-là par Voïnarovski au directeur de l'Institut d'Électrotechnique Nicolaï Pissarevski, ont été conservées. Dans ces lettres, on trouve des détails intéressants sur l'organisation du processus d'enseignement dans l'Institut Montefiore il y a cent ans.

Paul VOÏNAROVSKI, étudiant de l'Institut Montefiore.

Dans la lettre du 11 octobre 1894 il écrivait : " J'ai demandé au directeur de l'Institut Montefiore de m'admettre dans le département des ingénieurs de cet institut … J'ai reçu une notification du recteur de l'Université de Liège et de Monsieur Gérard qui m'informe que grâce à mon diplôme d'ingénieur télégraphe et en particulier à mes diplômes français (ceux d'un licencié des Sciences Mathématiques), ainsi qu'à ma connaissance du français, j'ai été admis sans aucun examen en qualité d'élève assidu et au même titre que ceux qui ont terminé les instituts supérieurs en Belgique et que je peux donc me présenter à la fin de l'année aux examens d'ingénieur électricien ".

Malgré son jeune âge, Paul Voïnarovski était déjà à l'époque un excellent spécialiste. Il avait plusieurs projets dont celui de l'électricité à l'exposition française de Moscou (1891) ; il devient membre perpétuel du groupe électrotechnique de la Société Impériale Russe de Technique (société fondée la même année de sa naissance).

L'Institut Montefiore a joué un rôle important dans la formation des cadres électrotechniciens d'Europe et de Russie grâce à son cadre magnifique des professeurs, à la méthodologie novatrice de l'enseignement et à la bonne organisation des travaux pratiques. Les cours intensifs à l'Institut permettaient d'acquérir en peu de temps un grand nombre de connaissances. Les lettres de Voïnarovski nous renseignent sur le programme des cours, ainsi que sur la méthodologie et les travaux pratiques.

En plus de Paul Voïnarovski, soixante-dix autres ingénieurs furent admis. Le programme intensif comprenait des cours de 9 heures à 19 heures avec une interruption de 2 heures. Les conférences traitaient de sujets variés : la théorie du magnétisme et de l'électricité, l'induction électro-magnétique, les mesures électriques et magnétiques, les mesures spéciales ainsi que les mesures photométriques. Comme on peut le remarquer, l'accent était mis sur les mesures. Sur l'ensemble de ces disciplines, il fallait passer quatre examens.

Lettre du 4 décembre 1894 : " J'ai passé l'examen à l'Institut Montefiore sur le magnétisme et l'électricité statique. Les résultats finaux montrent que j'ai obtenu la meilleure note et que jusque-là on me considère comme le premier dans cette discipline. En ce qui concerne les connaissances pratiques, je m'occupe de la préparation d'un galvanomètre Deprez d'Arsonval et d'un électrodynamomètre. J'ai terminé ces appareils, il ne me reste plus qu'à assembler les différentes parties. Les cours au laboratoire prenaient 7 heures par jour, en plus de trois autres cours (3 fois par semaine, théorie du magnétisme, électricité, électromagnétisme, induction électromagnétique, mesure électrotechnique) ".

Les travaux pratiques dans les laboratoires étaient toujours des investigations indépendantes. Au début il fallait travailler sur la théorie d'un problème puis sur sa réalisation pratique. Pour ces travaux, il fallait choisir plusieurs questions scientifiques. Voïnarovski a réalisé les travaux suivants : " Recherche sur la perte de charge à travers la conductibilité diélectrique dans les condensateurs " et " Recherche de galvanomètre balistique ". Les méthodologies les plus nouvelles furent introduites dans l'enseignement. Par exemple en écrivant sur les projets de machines à courant alternatif et sur les engins polyphasés, Voïnarovski disait que ses projets étaient " très intéressants parce que plusieurs d'entre eux avaient pu donner naissance à des machines à courant alternatif ainsi qu'à des engins électriques polyphasés et ceci grâce à l'utilisation de méthodologies nouvelles élaborées par l'assistant de Bast sous la direction de Monsieur Gérard ".

Le programme des travaux pratiques était complété par une inspection des usines de Belgique, des stations locales et des machines à vapeur pendant leur fonctionnement. D'après une lettre de Voïnarovski sur la pratique à l'Institut Montefiore : " conformément à l'opinion de plusieurs personnes est organisée même mieux qu'aux autres institutions pareilles. En ce qui concerne d'autres observations, il me semble que le rôle principal ici est joué par une bonne sélection des laborantins et du chef d'atelier. Une grande partie d'appareils nécessaires pour les mesures de laboratoire est fabriquée à l'Institut lui-même selon les indications du laborantin. Donc, l'Institut a créé au fur et à mesure et par ses propres moyens, si pas tout le matériel, en tout cas, une majeure partie. Autrement ce n'est pas possible, car il faut avoir beaucoup d'appareils difficiles à acquérir vu les impératifs et les conditions spéciaux dont ces appareils doivent satisfaire pour l'éducation ". Et ensuite, sur les méthodes de l'enseignement du professeur Gérard : " Au cours de monsieur Gérard il y avait également des étrangers. Monsieur Gérard donnait son cours d'électrotechnique à la faculté d'ingénieurs de manière assez rapide, en expliquant en détail les problèmes les plus importants ". Pour d'autres sujets, il renvoyait au manuel qu'il recommandait pour l'étude d'une façon indépendante. Il est à souligner spécialement que les cours sur les mesures électrotechniques faits par le professeur Gérard à l'Institut Montefiore n'étaient pas édités par lui. Chaque étudiant faisait ses propres notes de ce cours. Et seulement en 1898 en Russie le cours intitulé " Mesures électriques " a été préparé pour la publication et édité en russe par Paul Voïnarovski selon ses notes et avec ses compléments. Le livre contenant 417 pages de volume et 255 dessins a été accepté à Saint-Pétersbourg comme manuel de référence pour les étudiants de l'Institut d'électrotechnique.

A la fin des études à l'Institut Montefiore, pour obtenir le titre d'ingénieur en électricité il fallait réaliser un projet et passer un examen annuel. Ainsi tous les éléments précis pour l'éducation des spécialistes en électrotechnique de haute qualité étaient inclus dans le programme des cours.

Paul Voïnarovski a pu se perfectionner comme scientifique et pédagogue grâce à l'éducation reçue à l'Institut Montefiore et aux conversations avec le professeur Gérard. Par ailleurs, ils sont devenus amis pendant la durée de ses études. Le professeur Gérard a remarqué l'élève Voïnarovski, qui était le premier de sa section. Il manifestait de grandes connaissances et un grand intérêt dans les domaines de l'électrotechnique ; ils discutaient souvent de questions scientifiques et conversaient sur les nouveaux records en électrotechnique comme le montre cet extrait de la lettre de Voïnarovski : " Je parlais avec M. Gérard et il n'y avait jusque-là pas de rapports détaillés sur ses expériences (il s'agit du transfert téléphonique par câble subaquatique). Néanmoins, comme le croyait Gérard, Precce ne tardera pas à communiquer les données détaillées, et moi, je consultais les journaux anglais… Ces expériences étaient réalisées en Écosse par Precce et étaient la continuation d'une expérience qu'il avait réalisée il y a deux ans ". Le professeur Gérard a aidé Voïnarovski dans ses recherches sur les fils électriques en Europe et lui a donné des lettres de recommandation pour les usines de câble en Allemagne et en Suisse.

Эрикъ Жераръ,

Директоръ Электротехническаго Института Montefiore.

ЭЛЕКТРИЧЕСКІЯ

ИЗМѢРЕНІЯ.

(Лекціи, читанныя въ Электротехническомъ Институтѣ Montefiore
при университетѣ въ Лютихѣ).

ПЕРЕВЕЛЪ И ДОПОЛНИЛЪ

П. Д. Войнаровскій,

Преподаватель С.-Петербургскаго Электротехническаго Института, Телеграфный Инженеръ,
Инженеръ-электрикъ Института Montefiore.

Съ 225 рисунками.

Принято какъ пособіе въ Электротехническомъ
Институтѣ.

С.-ПЕТЕРБУРГЪ.
Изданіе Ф. В. Щепанскаго, Невскій пр. 34.
1898.

La page de titre du livre *Mesures électriques.*

Diplôme de fin d'études de l'Institut Montefiore délivré à
Paul Voïnarovski.

Pendant ses études, Voïnarovski examine en Belgique la question de la télé-communication sur grande distance — l'aide du professeur Gérard a été " très importante pour cette recherche " (lettre à M. Pissarevski). Après les stages de Voïnarovski à l'étranger, le professeur Gérard a soutenu son élève en lui donnant l'opportunité de passer un examen annuel pour le titre d'ingénieur en électrotechnique. Voïnarovski a terminé brillamment ses études à l'Institut Montefiore.

Pendant ses études il a voyagé dans plusieurs villes d'Europe afin de connaître les constructions électrotechniques. Dans la lettre d'avril 1895 il écrit : " J'ai visité Bruxelles à Pâques et regardé un bâtiment neuf de la poste et télé-graphe, ainsi que son équipement. J'ai réussi à observer le chemin de fer élec-trique, l'usine de lampes et la station centrale de téléphone. Pendant les heures de loisir je découvrais la signalisation des chemins de fer belges ".

Après son retour à Saint-Pétersbourg, Voïnarovski continue à s'intéresser aux nouveautés dans les domaines des techniques européennes, la pédagogie des ingénieurs et à entretenir ses relations avec l'Institut Montefiore et son directeur M. Éric Gérard. Paul Voïnarovski est resté membre de l'Association des ingénieurs électriciens de l'Institut Montefiore jusqu'à la fin de sa vie. Mais ce qui est beaucoup plus important, c'est que l'expérience et les principes pédagogiques que Voïnarovski a acquis à l'Institut Montefiore ont été assimilés et développés en Russie, à l'Institut Electrotechnique de Saint-Pétersbourg.

Ces mêmes principes constituent la base de l'enseignement de l'Institut Polytechnique de Saint-Pétersbourg.

Élève de l'Institut Montefiore, de la Faculté des Sciences de l'Université de Marseille, de l'Institut Electrotechnique de Saint-Pétersbourg et ingénieur russe scientifique et pédagogique, Voïnarovski occupe une place proéminente dans l'histoire du développement de l'électrotechnique. Il a consacré 20 ans de sa vie à cette science en abordant différents aspects. Son oeuvre compte beaucoup de points innovateurs et professionnels. Il a donné des cours aux étudiants de l'Institut Electrotechnique, cours complexes et absolument nouveaux, et pendant 7 ans il a été le directeur de cet institut après la mort d'Alexandre Popov, le célèbre inventeur de la radio. Voïnarovski rêvait de transformer son institut supérieur en un institut exemplaire tout en s'inspirant de l'expérience de Montefiore.

Dans sa note au Ministre des Affaires Intérieures de 1910, il écrivait : " Compte tenu de l'évolution de la vie politique, l'industrie et la technique sont deux facteurs importants de la croissance, j'ose donc penser qu'il est grand temps et nécessaire de reconsidérer sérieusement la question de réorganisation de l'éducation technique en général et de l'éducation technique supérieure en particulier ". Et son rêve fut réalisé. Dans le nouveau bâtiment de l'institut, grâce à l'équipement des laboratoires, toutes les conditions étaient réunies pour préparer les futurs ingénieurs au niveau théorique et pratique. Le professeur Voïnarovski a introduit beaucoup de nouvelles disciplines dans les instituts russes, dont la transformation de l'énergie sur une longue distance (1892-1893), la télécommunication et les signalisations des chemins de fer (1893-1894), les mesures électriques, la transformation de la distribution de l'énergie, les réseaux électriques et les fils électriques. Les manuels des cours constituaient une sorte d'encyclopédie de l'électrotechnique de la fin du XIXe siècle et du début du XXe. Son souci était la création d'un potentiel scientifique technique qui attirerait les idées les plus représentatives de l'électrotechnique en Russie et permettrait d'établir des liens avec les plus grands centres électrotechniques mondiaux. On peut nommer plusieurs noms qui sont liés à l'enseignement de cet institut à cette époque : Iakov Hackel, Henri Graffetio, Stéphane Timochenko. Les élèves de Voïnarovski sont devenus les meilleurs électrotechniciens en Russie et l'institut, un centre de recherche en électrotechnique.

L'activité polyvalente pratique et scientifique dans les différentes sphères de l'électrotechnique, de la télécommunication et les tramways électriques à courant puissant et haut voltage, est tout aussi importante que son activité pédagogique. Il n'y avait aucune question concernant l'électrotechnique qui soit restée sans réponse : le développement des questions scientifiques appliquées, l'activité des ingénieurs et des experts, le développement des premières règles de sécurité électrique et de réception des constructions électrotechniques, l'organisation et la participation active à tous les congrès nationaux et internationaux.

Le 25ᵉ anniversaire de l'Institut d'électrotechnique de S. M. I. Alexandre III à Saint-Pétersbourg. Le premier directeur de l'Institut, N.G. Pissarevski et P.D. Voïnarovski,devenu directeur de l'Institut le jour de cet anniversaire. La façade du bâtiment. (Photo de la revue *Niva,* n° 1 (1912))

Le nom de Voïnarovski est lié au projet de la ligne téléphonique entre les deux capitales Moscou et Saint-Pétersbourg : la plus longue à cette époque-là en Europe, réalisée en 1898. En 1903 à l'Institut Electrotechnique, Voïnarovski créé en Russie le premier laboratoire de haut voltage (200.000 volts). On a expérimenté l'isolation et le matériel du réseau de tramway de Saint-Pétersbourg et examiné les déchargeurs produits par Siemens, Land und See-Kabelwerke, Schukkert Werke et Westinghouse.

Il a écrit des oeuvres importantes : *Les lignes électriques souterraines* (1910) et *La théorie du câble électrique* (1912, republié en 1921). Elles ont constitué les bases de la technique des câbles. Le livre *Théorie des câbles électriques* est resté durant de nombreuses années un manuel essentiel pour les étudiants et pour l'activité pratique des spécialistes. Dans cet ouvrage sont abordées toutes les questions liées aux câbles : théorie, technologie, construction, moyens de mise en place, méthodes de calcul des régimes de température des câbles de puissance, atlas et schémas des méthodes de mise en place des câbles.

C'est étonnant, mais l'activité polyvalente et intensive de Paul Voïnarovski s'est réalisée dans une vie pourtant courte. Voïnarovski est décédé le 26 juin

1913 à l'âge de 47 ans. Sa sépulture se trouve à Sébastopol où il est né et a passé son enfance.

Nous exprimons notre profonde gratitude à ceux qui ont accordé l'aide financière permettant de réaliser le présent article : notamment aux Messieurs J.N. Delanaye et M.J. Jasson respectivement Directeur régional et directeur régional adjoint de la Société SPE (Société coopérative de production d'électricité) ; à Monsieur A.D. Victorov, président du Comité de tutelle du Fonds Transfert-Start (Saint-Pétersbourg).

BIBLIOGRAPHIE

Matériaux des Archives Historiques de Saint-Pétersbourg, fonds 990, inventaire 1, dossier 328.

THE MODERNIZATION OF CHINA'S MECHANICAL ENGINEERING UNDER THE INFLUENCE OF THE WEST (1581-1985)

Baichun ZHANG

FOREWORD

Mechanical engineering in ancient China has been described by some scholars. Here I would like to talk about the modernization of China's mechanical engineering under the influence of the West since the middle of the 19th century. My paper is divided into four parts :

I) Early introduction of western mechanical engineering (1581-1840).
II) The introduction of modern mechanical engineering (1840-1914).
III) An embryonic system of mechanical engineering takes shape (1914-1949).
IV) The establishment of the system of contemporary technology (1949-1985).

In brief, the history of technology in modern China has not been one of inventions, but one in which modern technology has been imported to China and implanted there. China was hindered in developing its basic industry and technology because of wars, political chaos, conservative ideas, ideological conflicts, and so on. In the 1950s, China massively imported technology and established its technology system and basic industry. However, its industry, which has lacked rational co-operation with research institutes and universities, has not achieved the ability to create varied advanced techniques, and thus its technology has basically remained at the stage where it merely copies foreign products.

EARLY INTRODUCTION OF WESTERN MECHANICAL ENGINEERING

The introduction of the western clock technology

Traditional Chinese mechanical engineering had basically come to being by the Qin and Han dynasties (221 BC-220). In the 11th century traditional technology matured to the point where complicated water-powered mechanical clock and astronomical instruments were developed. Yet, by the 15-16th century, however, the Chinese made hardly any mechanical clocks and knew virtually nothing about the development of western technology.

In the 1550s, Jesuits started their religious missions to China, but they were forbidden by the Ming Government to preach beyond Macao. In 1578, Alexandre Valigani, a Jesuit in charge of Indian and Far-eastern affairs, arrived in Macao and developed new strategies. He encouraged his missionaries to learn Chinese and local customs. In the following years, Michel Ruggieri and Matteo Ricci used clocks and prisms to amuse and flatter Guangdong governors, and as a result, were allowed to build a church in Zhaoqing. In Jan. 1601, Matteo Ricci came to Beijing and presented gifts such as clocks to Emperor Wanli of the Ming Dynasty, who then showed strong interest in European civilization. All emperors of the subsequent Qing Dynasty took a fancy to European timepieces. They invited European craftsmen to come to the royal workshops and make clocks upon their requirements. Royal clock-making reached its height during the reign of Emperor Qianlong (1736-1796). Although the chief craftsmen were Europeans, many Chinese were also involved, blending Western techniques with China's traditional arts. European clocks were first copied between 1620 and 1630, and some individual clock producers in Suzhou and Guangzhou (Canton) became well-known for their works[1].

The introduction of western mechanical knowledge

After contact with the Jesuits, some Chinese believed that Western technology would benefit Chinese society. They began studying Jesuit scientific knowledge and worked with them to translate it into Chinese. In 1612, Xu Guangqi and Italian Jesuit Sabbathin de Ursis jointly published *Hydraulic Technology of the West*, giving detailed introductions to the Archimedean screw pump and piston pumps. In 1626, Wang Zheng asked Johann Terrez to help him translate practical mechanics into Chinese. A year later, they published *Diagrams and Explanations of Wonderful Machines*, which covers statics and dozens of mechanical structures. However, majority of the new knowledge was not used by the Chinese because traditional Chinese machinery satisfied the practical needs at that time[2]. In the late 1670s, Belgian Jesuit Ferdinand Verbiest demonstrated his small four-wheeled wagon driven by an impact steam turbine to the Qing emperor in Beijing, but this machine was not taken up by the Chinese.

---

1. Zhang Baichun, " The Importation of European Clock and Watch Technology into China and the Questions Related during the late Ming and Qing Dynasties (1580-1911) ", *Journal of Dialectics of Nature*, vol. 17, n° 2 (1995), 38-46.

2. Zhang Baichun, " A New Study on " Diagrams and Explanations of Wonderful Machines of the West " Written by Wang Zheng and J. Terrez in 1627 ", *Journal of Dialectics of Nature,* vol. 18, n° 1 (1996), 47-53.

In the 16th Century, Chinese scientists focused on the structure and functions of machines, while western scholars sought to analyse simple machinery with the aid of mechanical concepts and geometric theories, in order to proceed to a better understanding of complicated machines. Herein lies one reason for the future gap between Chinese and Western developments in engineering. While the Chinese might improve their mechanisms through their experience, Westerners began to combine their technology with science.

Before 1840, the introduction of Western science and technology into China depended mainly on activities of Western missionaries. However, missionaries came to China to preach their religious doctrines, and science and technology was but a key to China's closed doors. The Qing government expelled them from China in 1723, cutting its only link with Western technology for the following 100 years. China did not again witness modern technology until European armies came to China in the mid-19th Century.

THE INTRODUCTION OF MODERN MECHANICAL ENGINEERING

Response to modern steamships and guns

The first steamship appeared in Chinese waters in 1830, but it was the Opium Wars, which showed the Chinese how powerful British ships and cannons could be. Some Chinese tried to study this modern equipment. Lin Zexu, the Qing envoy sent to Guangdong to eliminate the opium trade, insisted on using European rifles and guns. Meanwhile, a well-known scholar, Wei Yuan, proposed that Chinese should " learn from advanced foreign technology and use it to beat foreign invaders ". However, their proposals were not adopted by the government. China had occupied a dominant position in East Asia for a long period, with the result that the Chinese felt superior in civilization and economy to other countries. The Qing Dynasty still held onto this historical sense of superiority and would not admit to its backwardness or learn from the " barbarous " foreigners. Although the Qing army lost the first Opium War in 1842, many Chinese still believed their traditional weapons were invincible.

In 1851, a large-scale rebellion against the Qing government broke out in south China. Two years later, rebels declared that Nanjing became capital of their Taiping Heavenly Kingdom.

It was not until China lost the Second Opium War and the Taiping Heavenly Kingdom controlled the greater part of south China in 1860 that it began to take modern guns seriously. In Jan. 1861, Zeng Guofan proposed to " learn foreign knowledge for making guns and ships " and was supported by both the court and other pro-western bureaucrats[3]. They suggested establishing a mod-

---

3. *The Whole Story of Arrangement of Diplomatic Affairs (during the Reign of Emperor Xian Feng)*, vol. 8 (1979), 2669.

ern military industry, but were strongly opposed by conservatives, who belittled the functions of western machines. However, because the conservatives had no way of countering domestic rebels and foreign invaders, their opposition was blunted. Finally, the government permitted adoption of limited Western technology to meet its urgent need.

The pro-western faction, however, stuck to the principle of " using Chinese theories as a core, and western technology for reference ". They refrained from copying civilian machines, saying it would undermine the social and economic order. After the Qing army was defeated by Japan in 1895, foreign goods poured into the Chinese market. The Qing government then had to declare " making domestic machines " as one of its national strategies. After 1898, it issued policies to encourage inventions and the copying of imported machines. Clearly, it was always when they had no other choice that the Qing authorities moved a step forward towards modernization[4].

The introduction of machines and early industrialization

In Dec. 1861, Zeng Guofan established the Anqing Arsenal, where, depending on traditional handicraft techniques, Chinese experts built a working steamship. In 1864, Li Hongzhang purchased his Suzhou Arsenal machines, which had previously equipped the British fleet.

In Nov. 1863, Rong Hong, returning from the US, advised Zeng Guofan to set up a " parent factory " and then sub-plants producing basic machinery[5]. Soon, Zeng sent Rong abroad to buy equipment. He and Li also established the South China Arsenal in 1865. Later, the Qing Dynasty imported machines and set up more than 20 factories engaged mostly in military production. One of them, the Fuzhou Shipyard, began producing military ships and steam engines from 1869. In 1883, Chinese engineers assembled China's first cruiser there. Another factory, South China Arsenal, made a Yangtze passenger steamship with 3000 horsepower, the best of that period. To train more people in Western technology, two factories opened schools. South China Arsenal also opened a translation institute.

The pro-western faction focused their attention only on weaponry production, and did not establish any basic industries or manufacture machines on a massive scale, so that the military industry continued depend upon imports and landed itself in a passive position.

Of course, a small number of people stressed the importance of machines in civilian production. In 1873, Chen Qiyuan reeled silk on modern machines. A year later, Li Hongzhang first proposed using machines to tap iron ore. In 1878, Zuo Zongtang imported machines from Germany to build a woollen

4. Zhang Baichun, *A Brief History of Modern Machinery in China (1840-1949)*, Beijing, 1992, 16.

5. Rong Hong, *My Life in America and China*, 1991, 83-86.

cloth factory in Lanzhou. However, society at large was reluctant to accept new things. Most of people were accustomed to outmoded customs, and traditional forms of production and old technology, and chaos of the times made development impossible. From the 1860s, small privately-run machine factories had emerged in cities like Guangzhou and Shanghai. When they repaired machines, they gradually learned how to reproduce some of them, following models to produce lathes and steamships. In the early 20th Century, they copied internal combustion engines. However, because China had lost sovereignty over its tariffs, domestic mechanical products were not protected, and private businesses could not compete with foreign manufacturers.

AN EMBRYONIC SYSTEM OF MECHANICAL ENGINEERING TAKES SHAPE

The changing distribution of mechanical industry

In 1911, the National Revolution broke out. Finally, the Republic of China replaced the Qing Dynasty, but the development of industry and technology was hindered by civil wars and the War of Resistance Against Japan before 1950.

China's privately-owned businesses in the mechanical industry improved during World War One, when Western powers sharply decreased their exports of machines and other commodities to China. From 1928, the National Government implemented a number of policies to encourage the private mechanical industry. In Shanghai and neighbouring coastal areas and the Yangtze River valley, some factories were in operation under the supervision of engineers, but they had only old machines and they were unable to guarantee the quality of their products.

In 1919, Sun Yat-sen suggested using foreign technology to set up state-owned machinery factories. His proposal had a direct impact on the industrial policy of the Republic of China. After Japan occupied Northeast China in 1931, the government initiated an industrial programme to strengthen national defense. *The Construction Plan of Heavy Industry* drafted by the Resources Commission in 1936 advocated importing manufacturing technology and building machine works in inland Hunan Province.

The mechanical industry suffered heavy losses after 1937 when the Japanese launched an out-and-out invasion against China. The National Government was forced to move some of these factories to the economically backward areas in the Southwest and Northwest. In 1938, the Machine Works of the Resources Commission moved to Kunming, Southwest China, but it could no longer maintain its planned production. Some important materials could not be imported during the war, forcing the factories to seek substitutes. Unfortunately, in 1946 the civil war broke out again. *The Construction Plan of Heavy*

Industry virtually failed to materialize, and private businesses could in no way turn the tide.

A basic grasp of copying techniques

After 1914, state-run factories started importing technology and assembling machines so as to achieve self-reliance. In 1920, the South China Shipyard made for the US a transportation ship with a capacity of 14,000 tons. Importing technology from Germany's Benz, the China Automobile Manufacturing Co. began assembling lorries in 1937. It even made engines by itself during the anti-Japanese war. Private factories, meanwhile, also improved their copying techniques. For instance, Xinzhong Co. made internal combustion engines based on models. Unfortunately, production failed to grow due to the destruction of the war.

The National Machine Works of the Resources Commission was the largest of its kind during the war. Its products included machine tools, water turbogenerators and internal combustion engines. The production of the countless private factories was small-scale, but they could copy many kinds of machines to replace imports.

By 1950, China had been able to copy ordinary machine tools, power generating machines of low power, and other machines. However, most of them were unable to compete with imported items in terms of power, precision, functions and durability.

The establishment of educational and academic institutions

During the late Qing and Nationalist period, sporadic attempts were made to develop education and academic institutions.

From 1872, the Qing authorities sent some 220 students to the US and Europe to study new technology. In the early 20th Century, more students went abroad to major in engineering. This modern education broke down the traditional Chinese belief that scholars should distance themselves from technology. Again, we see that was conservative attitudes and other factors prevented the development of a sound educational basis for mechanical engineering, industry and research.

In 1895, the Zhongxi College in Tianjin started a mechanics course, the first in China's higher-learning institutions. The Qing Dynasty worked out *The Platform of Universities* in 1903 to unify its school systems and curricula. In 1921, both the Communications University and the Southeast University initiated the departments of mechanical engineering. After that, such courses became widespread in China, with most of the lecturers being Chinese and textbooks in foreign languages. By Oct. 1936, 17 universities had opened such courses and 1,500 students had graduated. Teachers wrote textbooks in Chinese, Professor Liu Xianzhou compiling *Terminology in Mechanical Engineering* in Chinese, unifying the terminology for the first time.

During the war against the Japanese, major universities were moved to inland areas, and mechanical engineering received more attention. The Central University enrolled 3 graduate students in mechanical engineering from 1939 to 1950. To build the post-war economy, the National Government sent a large number of college graduates to the USA and the UK to further their studies in universities or factories. Most of them returned around 1949 to lead the country's technological development.

In 1912, China's first engineers' society was founded under the auspices of Zhan Tianyou and other scholars. In 1936, the China Mechanical Engineering Society was born. Engineers first made experiments on materials, internal combustion engines and substitute fuels in some simple laboratories of the National Bureau of Industrial Research in the thirties. During the war against the Japanese, they moved their labs from Nanjing to Chongqing, and then built their own factory.

THE ESTABLISHMENT OF THE SYSTEM OF CONTEMPORARY TECHNOLOGY

After a four-year civil war, the People's Republic of China replaced the Republic of China in 1949.

In the fifties, Western policies of isolation forced China to import technology only from the Soviet Union and East Europe. China managed to set up systems in research, design, manufacture and education, and was able to produce the majority of the machines it needed on its own. However, in the 1958 " Great Leap Forward " movement, the country blindly pursued the quantity of grain, steel and machine production, ignoring their quality. Following the breakdown of Sino-Soviet relations in the early 1960s, China depended only on itself to develop its science, technology and industry. After the start of the " Cultural Revolution " (1966-1976), both the professional training and development of manufacturing technology came to a halt. Under the slogan of " opposing the worship of foreign things ", China stopped its imports of advanced technology. As a result, its development of mechanical engineering lagged even further behind the West.

The development of research in mechanical engineering

The government of the P. R. China set up a planned economy after 1949, which included the development of science and technology. In 1952, universities and colleges followed the Soviet model and set up different majors in mechanical engineering. Four years later, China made *The Long-term Program for Science and Technology between 1956-1967*. It covered the development of precision machines, jet engines, etc. The First Ministry of Mechanical Industry

set up a series of academies and institutes, including the Academy of Machinery Science and Technology, either on its own or together with universities.
Some major enterprises also had their own research units. In 1958, China and
the Soviet Union signed an agreement under which the latter promised to
either work with or aid China in major scientific research projects, including
the design and manufacturing technology of large-scale equipment, precision
machines and precision instruments. From 1950 to 1965, China sent more than
16,000 students to the Soviet Union and East Europe.

China started research in mechanisms and mechanical transmission in the
1950s. The research aimed primarily at developing new products. In the sixties,
research centred on large sets of high precision, high quality and high technology equipment, and digesting the technology imported from the Soviet Union
and East Europe.

Owing to some injudicious decision-making, and the huge demand of the
market, the mechanical industries paid great attention to products and equipment, with the result that basic technology and machining techniques remained
backward. Although some research made early advances, it took years for
results to be put into use. For example, numerical-controlled machine tools
were first made in 1958 and marketed in early 1960s, but they had not been
popularized by mid 1970s. Meanwhile, research in some important spheres,
such as superhigh precision machining and mechanical liability, was not
started until the seventies.

Since China adopted its policies of reform and opening-up, it has established a number of new laboratories which train graduate students and conduct
research in some pioneering subjects. Some research has come up to advanced
world levels. However, there still exist many problems. For example, experi-
menting means are backward, original and creative research and research based
on huge amounts of data are still insufficient, and manufacturing industry still
lacks some necessary technology.

The achieving of designing capability

In the early 1950s, China mostly copied Soviet products, including their
design, machining technology and technological standard. When it used foreign technology, China gradually grasped some critical techniques and designing methods which were still centred around copying. It promoted the 3-phase
designing method which covered project design, technological design and construction work design. But such an experimental designing method was still
close to copying, analogy and magnifying.

After Sino-Soviet ties were cut, the Chinese mechanical industry transformed its basis of designing from mere experience to research, calculation and
analysis. It also increased the percentage of high-level products and complete
sets of equipment. By the early 1970s, China had succeeded in making high

precision machine tools, such large equipment as 30,000-ton hydraulic press, 300,000-kilowatt hydropower or thermal generators, complete sets of oil refining equipment and fertilizer production equipment. But the overall designing capability had yet to improve. By the late 70s, few machines made in the 50s had been updated. For instance, the technology used for the Jiefang lorry was only that of advanced nations in the 1940s.

Since the 80s, computers have been applied to mechanical designing. On the whole, however, designing upon experience rather than scientific data still prevails. In addition, designing standards are obsolete. Naturally, it is difficult to renew old designs in time. Among 50,000 mechanical products in 129 kinds, about 60% are found to involve pre-70s technology of developed countries[6].

Grasping of fundamental manufacturing technology

In the fifties, China speeded up its process of industrialization and focused on the development of state-owned industries. Meanwhile, it brought private businesses under government plan and transformed them gradually into state-owned assets. Under a 1952 agreement, the Soviet Union helped China build 141 industrial projects, which increased to 156 in 1955. Among them, 26 were related to the mechanical industry. In the latter half of the fifties, the mechanical industry focused on its attention on metallurgical equipment, power-generating units, transportation machines, metal cutting machine tools, etc., laying the foundation for manufacturing technology. Domestically made machines accounted for more than 60% of the national total[7].

In 1964, China started construction in the inland areas to avoid possible attack by the West or the Soviet Union. Machining technology only improved slowly because of backward research, the slow spread of achievement and the lack of technological imports. New projects followed the technology of those aided by Soviet Union in the fifties. By the late 70s, China's complete sets of equipment had lagged two or three grades behind foreign equivalents in terms of scale, parameter and production.

Since reform took root in China, the government has continued working on major research projects and introducing technology, and increased its ability to develop crucial equipment and complete sets of equipment. Meanwhile , the mechanical industry has equipped itself with new equipment, including numerical-controlled machine tools and industrial robots. By the late 1980s, 85% of China's mechanical products were made domestically[8].

With the increase of technological imports, both mechanical products and technology have improved. But on the whole, the introduction of technology is

6. State Natural Science Foundation, *Mechanics*, 1994, 75, 32 ; Jing Xiaochun, *Mechanical Industry in Contemporary China*, vol. 1 (1990), 39-60.

7. Jing Xiaochun, *Mechanical Industry in Contemporary China*, vol. 1, *op. cit.,* 39-60.

8. State Natural Science Foundation, *Mechanical Manufacture (Cold Working)*, 1994, 21.

still in the stage where it copies foreign technology and makes products which the West was already able to. Compared with advanced countries, Chinese products are less admired in technology, reliability, energy consumption and durability. They can last for 1/3 to 1/2 of the time of Western products. The exports of China's mechanical products occupy only 0.5% of the word total[9].

POSTSCRIPT

1. Technology and industry has not been able to develop smoothly in modern China because of the instability of society and the economy. Since 1840, this country has been repeatedly plunged into wars or political chaos. Its mechanical industry still can not meet the demands of national economic development and defence, and input into the industry has been limited. This has forced decision-makers to take measures which can bring immediate benefits, but ignore fundamental research and mastering all imported technology.

2. China has been following in the footsteps of advanced countries and trying to catch up with them. Generally, it has taken technology introduced from developed counties as a starting point for further improvement. Based on assimilating imported technology, an independent capacity for developing new technology can gradually come into being. Nevertheless, its development of technology has easily fallen into a vicious circle of " importing — lagging behind — reimporting " due to the lack of basic technology and basic industry over a long period.

3. A certain times, the introduction of the western technology into China was hindered or suspended. During the second half of the 19th century, China delayed total acceptance of modern mechanical engineering mainly because the conservatives strongly opposed learning from the West. In the 1960s and 1970s, China could not make use of the latest achievements of the West and the Soviet Union to improve its technology because its link with developed countries was cut by contrasting ideologies.

4. Research has been divorced from application. Universities and research institutes have focused on theories and experiments, while factories, where design was close to copying and experience, paid attention to production of machines. In addition, state-run factories and research institutes were inefficient. Since 1979, the government has been trying to set up a new system to combine the creative power of engineers, researchers and technicians with the imported technology.

9. State Natural Science Foundation, *Mechanics*, Science Press, 1994, 75, 32 ; Jing Xiaochun, *Mechanical Industry in Contemporary China*, vol. 1, *op. cit.,* 39-60.

A SHORT HISTORICAL AND EPISTEMOLOGICAL ANALYSIS OF THE SIGNIFICANCE OF THE WORK OF ROMANIAN ENGINEER AND PHYSICIST ALEXANDRU PROCA (1897-1955)

Liviu SOFONEA

Unlike many other researchers (theorists) Alexandru Proca[1] (1897-1955) started from *practice* (*i.e. engineering activities, experimental investigations*) to arrive to important theoretical discoveries and interpretations which put his name in the *Golden book of Physics*[2].

The representative scientific papers of this eminent scientist are :

" Rayons bêta du mezo-thorium ", *Comptes-rendus de l'Académie des Sciences de Paris,* 183, 878 (1926).

" Équation de Dirac ", *C. R. Acad. Sci. Paris*, 190, 1377 (1930).

" Réflexions sur la Dynamique 5ᵉ dimensionnelle ", *C. R. Acad. Sci. Paris*, 186 (1928), 739.

Sur la théorie relativiste de l'électron de Dirac en un champ nul, Thèse de doctorat en Physique, Sorbonne, 1933.

" Solution des équations de Maxwell pour le vide ", *C. R. Acad. Sci. Paris*, 197, 1725 (1933).

" Mécanique quantique des protons ", *C. R. Acad. Sci. Paris*, 198, 54 (1934).

" Particules qu'on peut associer à la propagation d'une onde ", *C. R. Acad. Sci. Paris*, 198, 643 (1934).

1. G. St. Andonie, *Istoria matematicii în România*, Bucureşti, 1965.

2. G. St. Andonie, *Istoria matematicilor clasice, aplicate în România (Mecanica şi Astronomia)*, Bucureşti, 1971 ; George St. Andonie, " Alexandru Proca ", *Gazeta matematică şi fizică*, vol. XI, 9 (Bucureşti, sept. 1959), 516-528 ; S. Bălan, Nicolae St. Mihăilescu, *Istoria Ştiintei şi Tehnicii în România : date cronologice*, Bucureşti, 1985 ; C.G. Bedreag, *Bibliografia fizicii române. Biografii*, Bucureşti, 1957 ; V. Marian, *Figuri de fizicieni români*, Bucureşti, 1969.

" Théorie du positron ", *C. R. Acad. Sci. Paris*, 202, 136 (1936).

" Équations fondamentales des particules élémentaires ", *C. R. Acad. Sci. Paris*, 202, 149 (1936).

" Sur la définition des champs électromagnétique par potentiels et moment magnétique de l'électron ", *C. R. Acad. Sci. Paris*, 202 (1936), 641.

" Théorie ondulatoire des électrons et positrons ", *J. Phys.*, 7 (1936), 347.

" Équation d'onde pour particules à spin unité ", *C. R. Acad. Sci. Paris*, 2 (1938), 356.

" Masse du méson ", *C. R. Acad. Sci. Paris*, 208, 884 (1939).

" Particules élémentaires et l'équation Klein & Gordon ", *C. R. Acad. Sci. Paris*, 210 (1940), 563.

" Intégrales premières dans la théorie du méson ", *C. R. Acad. Sci. Paris*, 212 (1941), 669.

" Mouvement du méson ", *C. R. Acad. Sci. Paris*, 212 (1941), 75.

" Théorie des particules matérielles, leurs spin ", *C. R. Acad. Sci. Paris*, 214, 60 (1942).

" Un nouveau type d'électron ", *Portuguese Physica*, 1 (1942), 59.

The relevant aspects of the scientific activity are :

1. In the very rich scientific activity of A. Proca[3] the ideas methods and physical consequences of Atomistics (experimental facts, Quantum Mechanics). Special Theory of Relativity, Theory of fields occupy a very important place.

2. In 1923 he has elaborated his first scientific paper : and this way just on the Special Theory of Relativity ; the paper was published in *Buletinul Societății Politehnice din România*. In his paper he deals with Analytical Mechanics, Relativist Mechanics (relativist mechanics of electron, spinorial mechanics), Statistical Electronics, Physics of Elementary particles, Nuclear Physics, Radioactivity, Structure of Matter, etc.

3. The theorist has proposed a " theory of motion " more general as the Special Theory of Relativity : in which the representation of the presence & motion of objects is made in the four co-ordinates of the space and time and the 5-th co-ordinates which is associated (*i.e.* is a conjugate variable) to the mass ; he arrives to the conclusions that the space-time is atomised, the action is atomised, and the light must have immanently (both) wave and corpuscular aspects.

---

3. G. St. Andonie, *Istoria matematicii în România*, *op. cit.*

4. In the doctoral thesis the logical *discours* has the following phases.

4.1. Discussion about Dirac's theory of the electron[4] : the author give a rational relativist formulation, expressing the relativist symmetry by introducing the proper time τ .

4.2. The prime integrals of the free motion of electron are studied : it is showed that new simple prime integrals of the motion can be obtained :

4.3. The mathematical/physical aspects of this equation/theory are examined and interpreted : the bilinear operators, the complete algebra of Dirac matrices :

4.4. The equation of Gordon is generalised.

4.5. The physical interpretation of the ternars operators and of the operators of Dirac are given from the examination of the generalised form of Gordon equation.

4.6. The discussed Dirac's theory is applied (in complete manner) in the relevant simple cases : the plane wave, the packet of waves.

4.7. The mathematical relativistic representation of the motion of electron in quantum regime is achieved using the derivative of the functions with respect the variable co-ordinate which is the proper time ($\hat{d} = \dfrac{\partial}{.\partial\tau}$; in the obtained relations the terms $A = m_o\tau$ and $B = e\tau$ are canonic conjugated magnitudes).

4.8 The electromagnetic field of an electron (microparticle which posses charge and magnetic momentum) is determined by the specific electrodynamic potentials (by derivatives) : 4-dimensional orto-potentials $(\vec{A}, \Phi)$ *(i.e.* conventional potentials) and 4-dimensional pseudopotential $\tilde{\vec{A}}, \Phi$ which possess opposite variance at the reflection $\vec{r} \to -\vec{r}$, with regard to orto-potentials. The relations fields-potentials are :

(1)
$$\vec{E} = -\left(\nabla\Phi + \frac{1}{c}\frac{\partial\vec{A}}{\partial t}\right) - \nabla \times \tilde{A}$$

(2)
$$\vec{H} = \nabla \times \vec{A} - \left(\tilde{\nabla}\Phi + \frac{1}{c}\frac{\partial\tilde{\vec{A}}}{\partial t}\right)$$

5. He had (theoretically) demonstrated the possibility of the existence (in theoria) of heavy quantons having a non-null mass, m_μ, bigger than of the electron, but less than that the mass of nucleon :

(3)
$$m_e < m_\mu < m_p \approx m_h o$$

---

4. The theory is based on a special wave equation : $\hat{D}\Psi = 0; \Psi$ is a specific mathematical entity named spinor. Some historical-epistemological-physical-mathematical aspects can be seen in the monographical article L. Sofonea, " Concepţia cuantică relativistă asupra electronului ", *Studii şi cercetări de Fizică*, vol. 3 (1968), 25-60.

In the years ΔT = 1940-1950 these (hypothetical) objects were named by many physicists " with different words "[5] and some competent scientists have considered that they must effectively exist *in re* because are the " careers " of the nuclear (strong/weak) interaction with very short range of action ; the optimistical scientists believed also that in short time " mesons " (3) will be discovered in laboratory, or observed *in natura*.

A. Proca has given his theoretical arguments in the same time with the Japanese Hideki Jukawa and independently.

The Romanian scientist has established effectively also the equations of mezonic vectorial fields : they are relativist invariant relations. These equations was immediately studied by Louis de Broglie (which named them the " Proca's equations "), and by other theorists.

The quiding ideas of the research brilliantly finalised by A. Proca was :

5.1 The analogy between :

a. the searched equations — with quantum (*i.e.* dualist) relativist character, and

b. the Maxwell Lorentz equations — which are the correct classical (*i.e.* non quantum) relativist field equations of the electromagnetic fields,

$$\hat{E}^{Q.R.}_{M.L.}$$

physical continuum entities to which is associated quanta of photons with null rest mass, $m_{o_\gamma} = o$, and (by this quantification : dualization) these equations are interpreted as classical-relativist-" dualized " equations :

$$\hat{E}^{Q.R.}_{A.P.}$$

(4) $$\hat{E}^{Q.R.}_{M.L.,m_{0\gamma}=0.} \quad \leftrightarrow \quad \hat{E}^{\ Q.R.}_{A.P.,m_{0\mu} \neq 0}$$

5.2 The searched equations must be adequate to represent the motion (*i.e.* propagation) of a new kind of quantons[6] with a non-vanishing rest mass, *i.e.*

---

5. This kind of physical objects (theoretical, real) were considered and named (differently) by some experts during the *Curriculum Historiae Physicae* :

a. In the period of historical - *Tempus* ΔT = circa 1930 - circa 1950 " heavy photons "/ " mezons " ; these objects were theoretical.

b. In the period of historical - *Tempus* ΔT = circa 1950 - circa 1958 " mezons "/mezons ; these objects were theoretical/real.

c. In the period of historical - *Tempus* ΔT = after circa 1958 " hadrons "/hadrons ; these objects were theoretical/real.

6. V. Novacu, *Istoria Fizicii*, Bucureşti, 1966.

to be correct equations of motions for these relativist micro-objects with $m_{o_\mu} \neq o$.

The form of these searched equation must be that which result from the replacing d'Alembert equation/operator with the Klein-Gordon equation/ operator.

(5) $$\hat{E}_{M.L.} ; \Delta - \frac{1}{c^2}\frac{\partial^2}{\partial t^2} = \quad \leftrightarrow \Delta - \frac{1}{c^2}\frac{\partial^2}{\partial t^2} - K^2 . ; \hat{E}_{A.P.}$$

The differential equations for the scalar vector potential of field are[7] :

(6) $$\Delta\Phi - \frac{1}{c^2}\frac{\partial^2\Phi}{\partial t^2} = -4\pi\left(\rho - \frac{K_o^2}{4\pi}\Phi\right)$$

(7) $$\Delta\vec{A} - \frac{1}{c^2}\frac{\partial^2\vec{A}}{\partial t^2} = -\frac{4\pi}{c}\left(\vec{j} - \frac{K_o^2}{4\pi}\right)$$

(8) $$K_o \equiv \frac{\mu_o c}{h} \neq o$$

i.e.

9) $$(\square - K^2)A_\mu^p\{X_{\mu\nu}\} = 0$$

with the gauge

(10) $$\frac{\partial A^p}{\partial x_\mu} = o$$

$\hbar = (h/(2\pi))$ is the rationalised Plank constant, c is the value of the velocity of light in a vacuum, μ_o is the rest mass of the particle carrying the interactions ; the spin : is $s = 1\hbar$. At that time such kind of relativist-quantum vector mezonical field was a hypothetical object.

The differential equations expressed in terms of fields are :

(11) $$F_{\mu\nu}^p \equiv H_{\mu\nu}^p = \frac{\partial A_\nu^p}{\partial x^\mu} - \frac{\partial A_\mu^p}{\partial x_\nu} \equiv A_{\nu,\mu}^p - A_{\mu,\nu}^p$$

$$\mu, \nu = 1, 2, 3, 4$$

(12) $$A_\mu^p = (\vec{A}, i\Phi)$$

(13) $$\frac{\partial H_{\mu\nu}^p}{\partial x_\nu} + K_o^2 A_\mu^p = 0$$

(14) $$K_o = \frac{2\pi\mu_o c}{h} \neq 0$$

(15) $$rot\vec{H}^p - \frac{1}{c}\frac{\partial\vec{E}^p}{\partial t} + K_o^2\vec{A}^p = (\vec{4}pj)/c \cdot \vec{E} = -grad\Phi^p - \frac{1}{c}\frac{\partial A^p}{\partial t}$$

(16) $$div E^p + K_o^2\Phi^p = 4\pi p\vec{H}^p = rot\vec{A}^p$$

(17) $$\frac{\partial A_\mu^p}{\partial x_\mu} = 0$$

7. *Journ. de Phys.*, 7 (1936), 347.

The last condition, *i.e.* the Lorentz condition (gauge) imposed to potentials, yields automatically : by taking the derivative of the third equation with regard to $\partial/\partial x_\nu$ and having in mind that the product of the symmetric tensor $\partial^2/\partial x_\mu \partial x_\nu$ with the anti-symmetric tensor $H_{\mu\nu}^p$ is zero. In usual Electrodynamics the Lorentz condition is a subsidiary relation introduced by the gauge invariance, namely to circumvent the indetermination in the choice of the gauge : for the " heavy photons ", — or for the " relativistic electrons " (as was the initial view of Proca : because the contemporary equations of Dirac introduced states with negative energies) — the 4-potentials are explicitly introduced in the first group of equations : thus the equations for the vectorial " mesons " do not possess the invariance with regard to the usual gauge transformations of this kind of potentials.

6. Applying the fundamental laws of the dynamics of a material point for the reversible processes he has established a mechanical function which have a comparable role with the entropy in Thermodynamics : this " mechanical entropy " is identical with the Schrödinger's action.

7. He observed that the equation which express the elementary action of a point-like body in terms of energy and momentum can be written as a scalar product between two 5-vectors which are momentum and an elementary displacement which express a connection between the form & content of the principle of minimal action and that of geodetic motion.

8. Studies in Relativist Quantum Mechanics and Electrodynamics :

8.1 He shows that Schrödinger's equation is not an universal wave equation but the equation of a particular motion.

8.2 In a study on the wave (*i.e.* quantum) relativistic Mechanics of positives and negatives electrons (*i.e.* positons and negatons) he starts from the wave equation of Gordon and establish an equation which gives in all cases a positive energy, in opposite with the theory of Dirac in which occurs the negative energies.

8.3 He applies the values of the energies, momenta and charges of the microparticles in motion in vacuum (*i.e.* without external fields) in the case of photons (*i.e.* to neutral microparticles, having the energy proportional with the frequency) ; he has operated an inclusion of Theory of Light (Photonic Optics) in the Quantum Mechanics.

8.4 He considered the problem of existence of a kind of particle having the rest mass null ($m_0 = 0$) — *i.e.* similar to the mass of photon, γ, — and the electrical charge equal with the quantum of electricity

$$e_{p\gamma e^-} \equiv e_{e^-}^p$$

and named this " ideatical object " the " pure charge ".

8.5 He has examined the spinorial-relativist wave equation of Dirac electron, and have showed :

a) The wave function of electron is composed not by 4 spinorial components but by 16 such components.

b) He discovered new prime integrals of this quantum equation of motion.

c) He give new interpretations for the Dirac's operators of Quantum Mechanics.

d) He try to explain the difference between electron and proton, clearly attested by the experience accepting the hypothesis that between total mass (M) and the electrical charge (e) exist a dependence : $M = f(e)$.

8.6 Studying the Maxwell equations in vacuum he made interesting considerations :

a) New solutions can be obtained (by using 4 solutions : which satisfies some conditions) which permit to express a connection between the Electromagnetic Theory and the Quantum Theory of Photons ;

b) He have studied the derivative operators with fractionate order and have used these mathematical entities in some calculations (in the frame of the Theory of Photons).

c) The description of the fields by the electric-magnetic potentials is not complete : because it exist also another solution of the Maxwell-Lorentz equations which satisfies the condition of relativistic invariance. In his work on the magnetic momentum of electron he propose the use of the orto-electrodynamic potentials, $(\Phi, \vec{A})$, and of the pseudo-potentials $(\tilde{\Phi}, \vec{A})$, i.e. mathematical entities which have different variance under the reflection of spatial co-ordinates.

8.7 In other works of Relativistic Mechanics he give :

a) a scheme of a classical theory of free particles,

b) indicate new relativist equations of the elementary particles,

c) stipulate that in the Theory of the structure of matter it is necessary to be considered the reciprocal transmutation of (relativist) microparticles.

8.8 He has studied (in theoria) the mass of the physical object $\tilde{P}$ compatible/ announced (in theoria) by his established equations : the mass/masses of the heavy (i.e. with $m_{o\tilde{p}} \neq 0$) vectorial (i.e. with the $s_{\tilde{p}} = 1.\hbar$) micro-particle/s, i.e. with :

$$m_{o_{e^-}} \angle m_{o_{\tilde{p}}} \angle m_{o_{p^+}} \approx m_{o_n} \angle m_{o_{H_1^1}} \qquad \text{(see footnote 5).}$$

8.9 He has made an accurate experimental research on the β decay mezothorium[8].

---

8. C.G. Bedreag, *Bibliografia fizicii române, op. cit.*

From the theory of Proca emerge *in aesentia* the conclusion that in Nature can exist a kind of (fundamental/derived) Force (*interactio*) : a short range force decreasing with the distance in an exponential manner :

$$\varphi_{static} \approx \frac{1}{r} e^{Kr} ; K \approx m_{o_{Q\hat{P}}}$$

with a vectorial character (described by the 4-potential vectors :

$$A_\mu = (\vec{A}, \varphi), \mu = 1, 2, 3, 4$$

with a quantum nature, *i.e.* carried by a boson (the quanton $\vec{P}$) with the spin $s = 1\hbar$ characterised by a vectorial (symmetrical) wave function.

8.10 The intense researches in Nuclear Physics and in Physics of High Energies/Elementary particles made after the Second World War, have showed that :

a) particles with masses m_μ , with the limitation $m_{e^-} < m_\mu < m_{p^+} \approx m_n$ exist : they was discovered firstly in the cosmic rays, and later in different collision processes put in evidence in laboratories (in accelerators, in emulsions, etc.)

b) but these " mesons " have an average life time too " long ", *i.e.* they can traverse the nuclear domain without being captured by the nucleons : they cannot be the *carriers* of nuclear interactions.

c) The principal carriers of strong & weak nuclear forces are the bosons pions, which have a rest mass whose value is intermediary between those of electrons and of nucleons (proton, p^+, neutron, $n°$; they are fermions charged/ neutral) ; thus they are (in this sense) " mesons ", but their spins are $s = 1\hbar$.

d) The microparticles with spin $1\hbar$ and vectorial wave function exist in the complex processes of generation & annihilation of elementary/very simple particles : weak interaction, strong interaction, resonances, muons.

Thus the whole importance of the Proca's theory of the vectorial microobjects (" heavy photons "/heavy photons/" mesons "/mesons/" hadrons "/hadrons (see footnote 5) was revealed undoubtfully later, few years after the death of the scientist[9].

The personality and the scientifical achievements of Al. Proca — born in Romania, educated in this country, formed here as engineer and scientist, permanently attached by elevated feelings to the culture & civilisation of Carpathean-Danubean-Pontean Space — are at present honoured in his home land : for his exemplary life, for his works of perennial value Alexandru Proca is considered the most brilliant, creative, fecund Romanian physicist theorist[10].

9. The *Imago* of the Physics of elementary particles/High energies developed in the second part of the 20th century is more complexe ; in this frame the considerations and the equations of Proca can be placed more expresively.

10. G. St. Andonie, *Istoria matematicilor clasice, aplicate în România, op. cit.* ; G. St. Andonie, " Alexandru Proca ", *op. cit.* ; S. Bălan, Nicolae St. Mihăilescu, *Istoria Ştiintei şi Tehnicii în România, op. cit.* ; C.G. Bedreag, *Bibliografia fizicii române, op. cit.* ; V. Marian, *Figuri de fizicieni români, op. cit.*

Main phases of Alexandru Proca's *curriculum vitae* :

1. He is born in Bucharest at the and of 19th century : is the son of a building engineer.

2. He has studied at the College *Liceul Gheorghe Lazăr*, in Bucharest in the real section.

3. In 1915 he was enrolled at the Faculty of Sciences of University of Bucharest, where he was student until 1916.

4. In 1917, after absolving the *Şcoala de Ofiţeri de geniu* (School for military engineers) and to the entry of Romanian Kingdom in the First World War he was incorporated in the Romanian Army and was sent to fight in several battles.

5. In 1918 he attend the lectures of the *Şcoala de Poduri şi Şosele* (School of Bridges and Roads) in Bucharest, which in 1920 became the *Şcoala Politehnică* ; this school affords him grant which permit him to make a study travel in United States of America.

6. In 1923 he has received the diploma of electromechanical engineer as " chief of a promotion ".

7. In 1923 he begins to work as engineer at the society *Electra* in Câmpina, in the Prahova area reputed by the rich oil fields.

8. In the same time he was assistant of the professor engineer Nicolae Vasilescu-Karpen at the *Politehnică*.

9. In 1924 he published the volume *Întrebuinţarea electricităţii în industria de petrol* (The use of the electricity in the oil industry) : this is the first Romanian work in which is stipulated the use of electricity in some technical activities developed in the oil industry.

10. In 1924 he has drawn up, together with the mathematician Ernest Abason (1897-1942) and I. Tănăsescu, *Buletinul de Matematică pură şi aplicată*, which became later *Bulletin de Mathématiques et de Physique pures et appliquées de l'École Polytechnique de Bucarest*.

11. In 1924 he was sent by the staff of the Society Electra to Paris for specialisation. In 1925 he was enrolled at the *Faculté de Sciences* of Sorbonne ; here has attend the lectures given by reputed scientists : Paul Langevin, Jean Perrin, Marcel Brillouin, Marie Curie, *et. al.*

12. In 1928 he brilliantly obtain his *licence en Physique*.

13. From 1926 he has passed to the *Institut du Radium* leaded by Marie Curie, who had appreciated him. In this reputed scientific establishment he has made experimental researches on the β-decay of Mezothorium.

14. In 1930 he passed to the *Institut Henri Poincaré* ; here he worked with the scientist Leon and Marcel Brillouin, and with Louis de Broglie.

15. In 1931 he translated in French the book *Principles of Quantum Mechanics of P.A.M. Dirac* (the translation was made in collaboration with I. Ullmo).

16. In 1933 he affirmed his doctoral thesis entitled *Sur la théorie relativiste de l'électron de Dirac dans un champ nul*, in front of a jury composed by Jean Perrin, Leon Brillouin, Louis de Broglie.

17. At the *Institut Henry Poincaré* he worked on the redaction of *Revue d'Acoustique* and on the new *Annales de l'Institut Henri Poincaré*.

18. In 1935 he travelled in Denmark, where he studied in the Copenhagen School of Niels Bohr.

19. In the period 1936-1939 he continued to make research activity in the domain of the theory of (elementary) micro-particles (high energies) : he has studied theoretically the properties of the masses of heavy " mesons " (at that time pure theoretical entities), alone, respectively (in 1939) in co-operation with the American physicist Samuel Goudsmid.

20. In 1939, after the break out of Second World War he has requested to the authorities of France to obtain the French citizenship ; the request was accepted.

21. During the German occupation of an important part of his new home land he has supported in France from the beginning, the civilian & military motion *La Résistance*.

22. Consequent with his political options he left France conducted by Pétain-Vichy-st regime ; and has taken refuge firstly in Portugal (where he has lived and worked in Oporto) and after in Great Britain.

23. After the Second World War he come back to France and take again the work at the *Institut Henri Poincaré* in Paris.

24. He was nominated *Directeur scientifique* at the *Centre National de Recherche Scientifique*, and chief of the seminar for theoretical Physics in the main University of Paris, the reputed *La Sorbonne*.

25. In 1951 he represented the *Académie Française des Sciences* to the General Assembly of the International Union of Pure and Applied Physics, which has developed its activities in Copenhagen.

26. In 1954 he was invited by important scientific authorities of Japan to make conferences : in this country he has accepted the invitation and in the same year he has given in Tokyo several conferences about the mezonic fields.

27. In 1955 a cancer of throat caused his death.

The professor engineer and historian of Mathematics St.G. Andonie who was colleague with the scientist in Politehnica, and which has described his life, has noted : 47 scientific memoirs and 6 didactic works.

Almost all activity of this scientist is dominated by the relativist & quantum ideas & methods : *i.e.* he has lived under the relativist and quantum paradigm, and in the same time has major contributions to the physical & mathematical applications of the basical ideas of Theory of Relativity and different types/levels of Quantum Mechanics.

A detailed epistemological analysis of the Alexandru Proca's work in Theoretical Physics is made by some Romanian researchers Nicholas Ionescu-Pallas, Iovitiu Popescu, Liviu Sofonea[11], Tibor Toro ; they have considered also the connections between Alexandru Proca's idea and :

a. the discovery in 1937 by C.D. Anderson and S.H. Neddermeyer in the cosmic radiation a micro particle which has the rest mass $m_\mu \approx 200\ m_{e^-}$ and the average life time of existence $\tau_\mu \gg 10^{-13}$;

b. the discovery in 1947-1948 made by C.F. Powel which observed in the cosmic rays a microparticle with the rest mass $m_{o\mu} \approx 270 m_e$ and the spin zero $s = 0\hbar$, this scalar boson has a short life time $\bar{\tau}_{\bar\mu} \approx 10^{-13} s$;

c. the theory of Hideki Yukawa ;

d. the *status* of the Theory of elementary particles (specially that of mesons, pi-ons, hadrons resonances, gauge theories in the historical periods *ca.* 1950 - *ca.* 1960 ;

e. the experimental discoveries made in the years 1961-1963 : the vector mesons of hadrons type ($\omega, m_o = 783$ Mev ; dipionic resonance $\rho, m_o = 760$ Mev, tripionic resonance mezons φ). These experimental discoveries are considered by the physicists and epistemologs as a turning point in the *curriculum* of research in the domain of the Physics of vectorial micro objects ; these discoveries represents an experimental confirmation of Proca's ideas : they are arguments about the real existence of the vector mesons.

The scalar potential $\varphi = \frac{1}{r}(e^{-kr}, K > 0)$ was applied by Nicholas Ionescu-Pallas and Liviu Sofonea also in Gravito-dynamics[12].

We consider that the potential $\varphi = \frac{1}{r}e^{-kr}$ must be, pertinently, named the potential Seeliger-Yukawa-Proca.

11. L. Sofonea, *Teoria Relativității și Civilizația & Cultura din Spațiul Carpato-Ponto-Danubian* (The Theory of Special & General Relativity and the Carpathean-Pontean-Danubean. Culture & Civilisation, Some relevant historical-epistemological remarks concerning the period 19th century-*ca.* 1980), Brașov, 1998.

12. " The idea/hypothesis : the gravitation is an universal interaction with very long " efficaciously " finite range. Physical, mathematical, epistemological aspects ", *Physical interpretations of Relativity Theory*, London (6-9 September 1996), 274-288 ; also L. Sofonea, " Geometry and Cosmology. A. Physical-Mathematical-Epistemological exegesis of a non-conventional cosmological model ", *Tensor, chigasaki*, vol. 57, 1 (Japan, 1996), 55-63.

THE BEAUTY OF STEAM LOCOMOTIVES
(ABOUT THE CO-OPERATION OF TWO CREATORS OF STEAM LOCOMOTIVES)

Karel ZEITHAMMER

The Czech steam locomotives are well known in the world of railways today. Their technical data are known as well, but their creators remain anonymous. That is especially true about the creators of their external features and their overall visage. The designers of steam locomotives in the Czech republic have scored remarkable successes in this particular field of industrial design. Specifically, in the Škoda-Works in Plzeň the technical designers have always worked in close co-operation with artists. In this paper, I would like to introduce the work of the leading figure of the external visual design of the Czech steam locomotives — academic painter Vilém Kreibich. Kreibich started by designing the external features of already existing locomotives, but after World War II, his artistic designs preceded the technical sketches of design engineers, and the engineers were forced to work within the bounds given by Kreibich's artistic design. This was very unusual in the field of locomotive building at the time. From this collaboration arose locomotives not only technically perfect but extraordinary beautiful, too.

Vilém Kreibich has had a very good opportunity to live near locomotives since his birth in 1884. His father was an engine-driver on the Bohemian West Railway. His family resided in a locomotive shed, which was most likely the source of young Kreibich's great interest in steam locomotives. When he was 17 years old, he started his studies at the Prague Academy of Art under the respected Prof. Hanuš Schweiger. Even before enrolling at the Academy, he painted the Gölsdorf's " Atlantic " express locomotive created for the Vienna - Prague Railway which was the most powerful of its kind on the continent. Ten years later, he painted another Golsdorf's express locomotive and his big artistic progress is apparent by comparing these two paintings.

During the First World War, Kreibich painted a general view of the biggest arms factory of the Austrian-Hungarian monarchy — the Škoda-Work in Plzeň

— and a series of pastel pictures of the interiors of the great work-halls. This pictures initiated his co-operation with Škoda-Works that has lasted almost forty years thereafter. Despite he was also a famous portrait-painter at the time, he dedicated more and more of his time and effort to the railway theme.

The year 1928 was a very important turning-point in Kreibich's life. He met a young engineer Vlastimil Mareš from the Ministry of Railways in Prague, who was the future chief of the Ministry's of Railway Design Department. Their meeting in 1928 was a beginning of a very close friendship that lasted for almost 26 years until Kreibich's premature death, and that had a permanent influence on the construction of steam locomotives in the Czech Republic. Kreibich had with Mr. Mareš the possibility to spend unforgettable hours in various locomotive sheds each week, made frequent journeys on drivers cabs of locomotives together, and this was like a technical school for Kreibich. Reciprocally, Mr. Mareš has visited Kreibich's studio very often, and the two discussed the importance and function of different parts of steam locomotives and their influence on the total arrangement.

Kreibich's recommendations were mostly of aesthetic nature while Mareš preferred the functional aspect of locomotives. During these meetings, Kreibich learned to intuitively design not only beautiful, but also technically sound parts of locomotives. Both gentlemen always agreed on the best solution to a design problem. In this time, Kreibich also painted the first famous large pastel- and oil-paintings of the express-locomotives of the Czechoslovak States Railways. Among the pictures excels the pastel " The Three Sisters " that depicts three new Škoda's " Pacifics ". Nobody knew to capture the atmosphere of steam-traffic so vividly as Kreibich.

The first practical result of Kreibich's and Mareš's co-operation was the new colour-scheme of the ČSD-express locomotives from the beginning of the thirties. The usual black was replaced with diverse colours. The green boiler with transverse stripes from brass, green deflectors, cab and tender with yellow lines and red wheels, that were the main characteristics of the Czech express locomotives in the period between the wars. In the end of the thirties, Kreibich participated in the creation of the new unified appearance of the locomotives of the 2nd ČSD generation.

The zenith of Kreibich's creative work lies in the last ten years of his life between the years 1945-1955. His best friend, Mr. Mareš, was at this time already the chief of the Design Department of the Ministry of Railways. Kreibich's creativity was finally given complete freedom. Already the first post-war express locomotive of the ČSD — the powerful " Mountain-Type " — marked a breaking point in the existing practice of locomotive design. The new locomotive was first designed by Kreibich and only afterwards did Škoda's engineers work out the technical aspect of the engine. The colour scheme was new as well. The pre-war green was replaced with blue, but the wheels remained red. Other types of the new unified line of locomotives with the same architecture followed soon thereafter. They all had all-welded boilers, combustion chambers, Nichol's thermosyphones, and mechanical stokers. Many others had three-cylinder engines, all axes mounted in rolling bearings and some others locomotives had the main and couplingrods fitted with rolling bearings, too.

The prototypes of the new 3rd generation locomotives were being prepared in late forties while the 2nd generation locomotives were still produced by Škoda. The outside appearance of the 3rd generation locomotives was Krei-bich's work as well. After the excellent Chapelon's experiences with com-pound locomotives of the French States Railways, the 3rd generation locomo-tives had the compound three cylinder engines, too. The top designers of the Škoda-Works and Mr. Mareš consulted some technical problems directly with Mr. Chapelon. At the beginning of the fifties, the first three prototypes were made with wheel arrangement 4 - 8 - 2. They served especially for testing. Although they had certain " children's diseases " their output was excellent : 3.200 HP with the load per axle 16 t only ! They differed from the locomotives of the 2nd generation in colour, too : blue was replaced by crimson. The entire new unified series of locomotives for the fifties was designed based on the experience with these prototypes. At the top of the new line stood a new pow-erful express locomotive with wheel arrangement 4 - 8 - 4 for the speed up to 100 miles per hour, and a new heavy freight locomotive 2 - 10 - 4 with even a booster. Both types had the same type of big boiler. Unfortunately, this 3rd gen-eration line came too late. The electrification of the main lines was in a full swing, and at the same time, the Cold War broke out with the Czech Republic remaining in the Soviet sphere of influence. The entire Czech economy was reorganised according to the new doctrines. Instead of the above-mentioned modem locomotives, Škoda started to build the universal middle heavy " New War Locomotives " in large amounts in the beginning of the fifties. The only information about the planned 3rd generation locomotives that remained to this date is the unfinished construction documentation and Kreibich's interesting and inspiring designs.

FIGURES

1. The last Kreibich's design : the express locomotive for fifties for the speed up to 100 miles per hour.

2. The famous Kreibich's " The Three Sisters " - from the left : 386.0 and two 387.0.

3. The Gölsdorf's express locomotive painted by Kreibich in 1911.

4. The first Kreibich'sdesign of 498.0 presented by him to Mr. Mareš.

Etrich — The Aeronautical work and Life of DSc h.c. Igo Etrich (1879-1967)

Hanuš Salz

On the basis of newly disclosed archives an attempt has been made to break fresh ground in the History of Aeronautics by giving a true to life portrait of DSc h.c. Igo Etrich and his career as an aircraft projector and aviation pioneer. The strict fact-finding analytical method enabled to throw light on hitherto dark corners and putative assertions so that inveterate mistakes, false dates or incorrect conclusions could be brushed aside and, eventually, set right. As to the general lay-out proportionate stratification of the technological as well as commercial, economic, organisational and last but not least human aspects has been aimed at. Thus in a detailed, sometimes even thrilling monography of the author the achievements of Igo Etrich representing an extraordinary, but most intriguing chapter in the history of aviation are carefully scrutinised under the severe *sine ira et studio* guideline.

Born on Christmas Day 1879 in the country town of Trutnov (Northeast Bohemia - Czech Republic) as elder son of Ignaz, a comfortably well-off textile industrialist. Since his teens Etrich Igo had shown ardent leaning towards aeronautical studies with special interest in the gliding experiments of the then leading European aero-researcher Otto Lilienthal (1896). One of his hanggliders, the *Sturmflügelmodell*, was bought for studying, but mounted on a solid undercarriage for starting it proved a total failure. Thereafter, seconded by his father Ignaz morally and financially and by the Austrian Franz Xaver Wels, engaged as technical engineer he took soon his own way on choosing the winged seed of the tropical plant *Zanonia macrocarpa* (now *Alsomitra macrocarpa Blume Roemer*) as ideal design for the supporting plane of his future aircraft. " Security " was Etrich's slogan and what this little natural flying wing precisely excels in its prolonged end extremely stable soaring. Artificial models of the *Zanonia* wing were built in progressively larger scales till in 1905. The idea of a self-stabilising aircraft, powered or unpowered, was mature enough to be applied for patenting. Patents were then granted jointly to Igo Etrich and Franz Xaver Wels in Austria and Hungary (each of them with its own Patents-Office), in the United Kingdom, France, Italy, Belgium and, later on, in Russia — the Austrian patent n° 23465 *Flugmaschine*, dated on 10 March 1906, being the most important. For reasons not clear enough — maybe through some fear of interference on the part of Prof. Dr Friedrich Alhborn in Hamburg (Germany) who as early as in 1897 in his treatise *Über die Stabilität der Flugmaschinen* (On the Stability of Flying Machines) had suggested the *Zanonia* as model for airfoil construction— no effort was made to acquire patent rights in the very promising territory of Germany.

Two years later, in October 1907 in Trutnov, an unpowered cockpit-glider called *Zanonia* was ready for trials and its flights with F.X. Wels piloting were first of their kind in the former Kingdom of Bohemia and implicitly in the old Austro-Hungarian monarchy as well. The longest flight-path of Wels in October 1907 in Trutnov measured more than 200 m.

Now, the development of powered aircraft came as the next logical step. Thinking about the future commercial exploitation both inventors left the remote provincial town of Trutnov and moved to the all-embracing Austro-Hungarian metropolis of Vienna. Nevertheless, experiments with the powered glider on well undercarriage dragged on, the desired man-carrying aeroplane seemed to be far away. Strongly criticising each other and Wels unexpectedly pressing for a biplane construction of his own idea e. g. without the patented wing-form Etrich and his associate had a bitter parting. Nevertheless, their joint ownership of the patents in Austria (n° 23465, n° 51064) and elsewhere entangled them in a vicious circle of complicated legal fighting until Etrich paid out his former companion and employee in 1911. After all, it was the only way for him to get rid of Wels because also the second Austrian patent n° 51064 *Tragfläche für Flugmaschinen* (Supporting Plane for Flying Machines), dated on 11 December 1911, applied for solely by Igo Etrich, was declared contingent upon the prior Etrich-Wels patent n° 23465.

Etrich, clinging adamantly to his idea of self-stabilising aircraft and having found a most skilful foreman and reliable all-round pilot in his compatriot Karl Illner, constructed afterwards the experimental monoplane Etrich I powered by the French 50 HP engine of P. Clerget. But the very breakthrough did not turn up until February 1910 when the monoplane Etrich II made its maiden flight on the Steinfeld airfield in Wiener Neustadt (Austria). Named *Taube* (Pigeon - Dove) by virtue of its silhouette against the sky this Etrich II became the best-known far-famed design of Igo Etrich and an everlasting flying memorial to him. A hangar-atelier was rented in Wiener Neustadt for further envisaged aeroplane fabrication.

Within a couple of months the Etrich-Taube dominated the newly organised military airforce in Austria-Hungary and Germany while this country with its unyielding policy of " no military aircraft from abroad " posed a special problem to Etrich. Also in other European countries inclusive the United Kingdom and in the USA the Taube-type monoplane stirred up vivid interest. At this moment it was undoubtedly imperative to commence serial production without any delay. Overwhelmed with offers for co-operation Igo Etrich waved aside his own investments and decides rather to sell his patents or patents pending, first of all in the two most important countries : in Germany to E. Rumpler *Luftfahrzeugbau GmbH* Berlin, in Austria to *Motor-Luftfahrzeug-Gesellschaft mbH* Vienna (MLG). Here, of course, the efficient Austro-Daimler and the German Daimler-Mercedes engines powering the better part of Etrich-Taube aircraft should be duly appreciated.

Several aeroplanes were then delivered by MLG for the Italian army, the British navy and a Russian air-association, one was partly constructed for a French private person, one was built on special license in France. North Africa sought for the first time one of the Italian Etrichs as warplane in the 1912/1913 Libya conflict against Turkey. The Rumpler company was not allowed to export Etrich-Taube types.

The serial production of the Taube aircraft having been arranged for Igo Etrich returned to Trutnov and in 1911-1912 two compelling prototypes were built there in the air-department of his textile works and tested on the nearby airfield in Josefov : two-seater Etrich-Schwalbe (E. Swalow) and three-seated Limousine. The Limousine should be the first cabin aircraft in the world Etrich claimed, but in the history of aviation it is better noted for its lateral control by tilting wings.

Oddly enough, as hinted at before, Igo Etrich had made no attempt to patent the Taube plane configuration until June 1910 when closing contract with Rumpler Company. Not so oddly the *Kaiserliches Patent-Amt* in Berlin turned down the application later in 1911 objecting the original Etrich-Wels patent n° 23465/1906 in Austria. The disastrous consequences of this ruling were not long coming. Every aircraft manufacturer in Germany could copy — and indeed copied — the highly successful Taube model without any obligation to Igo Etrich, and even the contract-bound Rumpler Company refused to pay license dues by giving as a pretext some constructional novelties of their own. Etrich took legal steps against Rumpler with no other effect for the claimant, however, than the annulment of Etrich-Rumpler contract of June 1910. Moreover in Austria the implementation of certain license provisions was disputed by *Motor-Luftfahrzeug-Geselleschaft* as well as militaries grumbling over the MLG monopoly and bad execution of state orders. To settle this additional agreements had to be negotiated on either side : with MLG about the exploitation of Etrich's patents in the United Kingdom, France, Italy, etc., while the Austro-Hungarian Ministry of War, without any ceremony, began to build Etrich-Type aircraft in the army Aero-Factory on its own account.

It was, however, Germany with its expanding, now free-for-all market which represented real threat and serious challenge to Etrich's plans. Actually, no other choice was left to him except to quit as loser or to wage an aviation enterprise of his own, on German soil of course. For this purpose the two-men company *Etrich-Flieger-Werke GmbH* was founded with Igo Etrich and his father Ignaz as partners and a brand-new, though rather small factory erected near Trutnov, but across the German frontier in Liebau, Upper Silesia (now Lubáwka, Poland). Since autumn 1912 standardised types of Taube aircraft were constructed in Liebau while the affiliated *Sport-Flieger-Gesellschaft mbH* in Berlin was responsible for sales, marketing, pilot training, participation in prize contests, etc. A limited manufacturing and trading license was given to Gustav Otto *Flugzeugwerke* in Munich (Bavaria). In March 1914 Igo Etrich became the main co-founder of the much larger aircraft works *Brandenburgische Flugzeugwerke GmbH* in Briest near Berlin (1915 *Hansa-Brandenburgische Flugzeugwerke*). In spite of the fact that *Brandenburgische Flugzeugwerke*, *Etrich-Flieger-Werke* Liebau and *Sport-Flieger-Gesellschaft* Berlin were merged the Taube monoplanes disappeared from the production list to clear the way solely for biplane types designed by the technical director Erst Heinkel, in course of time originator of the first ever rocket-plane HE-176 and first ever turbo-jet HE-178 (both in 1939). Shortly after the World War I (1914-1918) had broken out, Igo Etrich sold his shares in the *Brandenburgische Flugzeugwerke* Briest and withdrew from the aviation business altogether.

As a distinguished episode in the Liebau period the so-called Five-Countries-Circuit (German/Berlin - Belgium/Brussels - France/Paris - England/Hendon Airport, London - The Netherlands/Nijwegen - Germany/Berlin) in September 1913 should be cited. In accomplishing this three primacies were scored : (1) Alfred Friedrich as pilot and his passenger Dr Hermann Ellias were the first German two-man crew to fly from Berlin to Paris or vice versa ; (2) the used Etrich-Taube monoplane was the first aircraft in Germany in operation with registration number viz. D-2, since the number D-1 was allocated, but never used ; (3) for the first time the official British air-space entering permit for foreigners was issued to Alfred Friedrich, Igo Etrich as passenger (in the second leg of the Five-Countries-Circuit) and their Etrich-Taube D-2.

Once more, at the turn of the twenties and thirties, Igo Etrich ventured on an ambitious project to create a " Flying Minicar ". Coming back to his pre-war conception of the 1912 Limousine and its special lateral control device he built a light two-seater cabin monoplane with a modernised Taube-wing and 40 HP Salmson engine (design engineer Erich Hackel). Unfortunately, unable to win sufficient interest for this construction at home, unable even to operate it for himself and hardly satisfied with certain offers from abroad he was forced to give up his last attempts in aviation. But as good luck would have it the prototype of the 1929/1930 Sports Limousine (Model n° E.VIII) outlived the World War II and is now exhibited with pride in *Národní Technické Muzeu*m (National Technical Museum) in Prague.

Besides Prague rare original Taube aircraft can be admired in Vienna (*Technisches* Museum, Etrich II-Taube 1910), in Munich (*Deutsches Museum*, Etrich-Rumpler-Taube 1910), in Oslo (*Norsk Teknisk M*useum, Etrich-Rumpler-Taube 1912, rebuilt as hydroplane START in Norway), in Berlin (*Deutsches Technikmuseum*, Jeannin-Stahltaube 1913). In Owls Head (Maine, USA) the Transportation Museum finished in 1990 the construction of an excellent flying replica of the Austrian MLG-Lohner-Taube combined type 1912/1913, while the replica of the same Taube, built as private undertaking in Austria and flown only once in 1991, fell short of expectations and its fate is blurred for the time being. Nevertheless, the first Etrich-Taube replica, original Liebau-type 1913, with the registration marking D-EFRI was demonstrated by Alfred Friedrich in Germany in 1936 ; its damaged air-frame is deposited in the *Museum lotnictwa* in Cracow (Poland).

As early as 1909 the Etrich-Taube models were brought out (author Etrich's aide Paul Podgornik-Hermuth). Since that time the Taube modelling is steadily progressing all over the world in innumerable forms and applications, for various purposes.

Many rewards and honours were conferred upon the Taube pilots Karl Illner and Alfred Friedrich and particularly upon Igo Etrich himself for his unique achievements in aeronautics — first of all the title of Doctor of Science honoris causa of the Technical University in Vienna (1944). Like many people of repute he wrote his memories and left a rather short message for generations to come in which, among other things, his long-standing attachment to parapsychological phenomena is fully professed. The earthly pilgrimage of DSc h.c. Igo Etrich ended in 1967 in Salzburg (Austria) where he rests in a grave of honour bestowed by the municipal government of Salzburg.

Also Karl Illner (1935) has got a grave of honour in Vienna and long before at the height of his career a memorial in Horn (Austria). In 1963, on the occasion of the 50th anniversary of the Five-Countries-Circuit, Alfred Friedrich was decorated with the French *Médaille de l'Aéronautique*. Friedrich died 5 years later, in 1968.

As for many other noteworthy persons — pilots, designers, experimenters — connected with the Taube-type aviation all over the world let us remember here (alphabetically) :

Gustave Amand :   air displays in France (1911-1912)

Amelie Beese :   the first German aviatrix (1911), air contests in Germany (1911), constructor of some Etrich-type Taubes for her own training school (1912-1914).

Heinrich Bier :   highly successful Austrian pilot, prize-winner and record breaker (1910-1911), the only competitor from Central Europe in the Daily Mail-Circuit of England (1911), air displays in Bohemia (1911), the only holder of two Gold Sports medals of the k. u. k. Österreichischer Aero-Club (1911, 1914).

Stanko Bloudek :   M.Sci. (Eng.), chief designer in the Etrich-Air Depart-
                          ment Trutnov and the Etrich-Flieger-Werke Liebau
                          (1911-1913).

Karl Caspar :   1911 founder of the *Hanseatische Flugzeugwerke*
                          Hamburg producing Hansa-Taubes, 1915 merged into
                          Brandenburgische Flugzeugwerke Briest.

Hans F. Dons :   Norwegian Navy officer piloting the hydroplane
                          START (1912-1914).

Paul Fiedler :   pilot for Etrich in the Trutnov- and Liebau- period
                          (1911-1913).

Giulio Gavotti :   officer of the Italian Air Force, war pilot in North
                          Africa (1911-1912).

Hellmuth Hirth :   chief pilot of the Rumpler-Company 1911-1912, tech-
                          nical director of the *Albatros-Fluzeugwerke* Berlin
                          1913-1914 (Albatros-Taubes), winner of the Kath-
                          reiner-Prize of 50.000 Mark in June 1911 (long-dis-
                          tance flight Munich-Nürnberg-Leipzig-Berlin), winner
                          of the Berlin-Vienna air races (700 km) in June 1912,
                          author of the book *20.000 km aloft* (in German, 1913).

Rudolf Holeka :   a distinguished Austro-Hungarian military pilot of
                          Czech descent (1911-1918), later on the first General
                          of the Czechoslovak Air Force.

Onikichi Isobe :   officer of the Japanese Air Force, constructor of the
                          Taube-type aeroplanes in Japan (1912-1914).

Ferdinand Konschel :   Austrian pilot, the third in the 1.100 km Schicht-Cir-
                          cuit Vienna-Prague-Vienna-Budapest-Vienna in 1914.

Božena Laglerová :   the first aviatrix in the former Kingdom of Bohemia
                          (1911), now Czech Republic.

Antal Lányi :   air displays in Hungary and Slovakia (1912-1914).

Yu Yan Lee :   officer of the Chinese Air Force, trained by K. Illner,
                          air displays in China (1912).

Heinrich Lübbe :   Rumpler-pilot, excellent flying performances in
                          Argentina (1913).

Miecislaus Miller :   as army Austrian pilot n ° 5 (1910), pilot of the first
                          military Etrich-Taube in Austria, Austria's aviator in
                          the Berlin-Vienna air races (1912).

Franz Oster :   German pilot residing in China, constructor of an
                          Taube-type aeroplane, air displays in China and Sri
                          Lanka.

Franz Reiterer :   Etrich's pilot, successor to P. Fiedler and A. Friedrich,
                          long-distance flights Berlin-Copenhagen and Berlin-
                          Vienna, competitor in the Schicht-Circuit (1913-1914)

Lilli Steinschneider :   the first Hungarian aviatrix, contesting in Austria and
                           Hungary (1912-1914).
Karl Stohanzl :   Austro-Hungarian military pilot, 1912 long-distance
                           flight Wiener Neustadt (Austria) - Jihlava (Bohemia,
                           Stohanzl's native town).
Albert Ziegler :   many air displays in Transylvania 1913-1914.

" Mortal is the worker, live is the labour ", said thoughtfully a Czech poet
about the human lot. No doubt that the Etrich-name and Etrich-Taube will get
easily and well-balanced over the milestone of the Third millennium.

FIGURES

1. Igo Etrich in front of his world-famous monoplane Etrich II - Taube
(Wiener Neustadt, Austria, 1910).

2. Etrich-Taube 1911 Type - Military Prototype in flight
(Wiener Neustadt Airport, Austria, Karl Illner piloting).

3. Etrich-Limousine 1912 Type, in front of it the first Czech woman-pilot
Bozena Laglerova (Josefov Airport, Czech Republic).

4. Etrich-Limousine E.VIII 1929/30 on the Prague Airport Kbely (Czech Republic),
well-preserved and nowadays shown
in the National Technical Museum in Prague.

5. Flying replica of the Etrich-Taube 1913 Type NM-3 (Reg. n° D-2) -
Owls Head Transportation Museum, USA, 1990.

THE STATUS OF *HOMO TECHNICUS-TECHNOLOGICUS EMINENS* IN THE LIMITED SITUATIONS. THE CASE OF ROMANIA IN THE PERIOD *ca.* 1ˢᵗ DECEMBER 1989 - 1ˢᵗ 1990

Elena HELEREA and Liviu SOFONEA

A USEFUL DEFINITION OF THE CONCEPTS *CULTURE* AND *CIVILIZATION* BASED ON AXIOLOGICAL CONSIDERATIONS

We characterize (*in ratio, in verbum*) " the human condition " — *i.e.* Man — by the definitions :

Man has conscience (in the most relevant part of His life) ; We express — *more philosophico*— this definition by the syntagm *Homo Conscious* (an accepted pleonastic syntagm, enforcing the meaning) ;

Man has options : he makes " accounts ", he defines many axiological referentials ; We express — *more philosophico* — this qualification by the syntagm *Homo Aestimans atque Hierarchicus.*

These (articled) definitions are pertinent *in aesentia* :

a. Man's Destiny doesn't come true in the world of things — *i.e. in Mundus Rerum*[1].

b. Man's Destiny is not realized in the natural world — *i.e. in Mundus Naturalis*[2].

c. Man's Destiny is defined in the world of Values & Valuations — *i.e. in the Axios Universe*[3].

---

1. There were societies in which the process of " reification " has been intensified (e.g. the " communist society ").

2. Man is an animal, but his essence is defined by spirituality ; that is by the conscience, not by the animality.

3. This manner of thinking and its terminology can be put reasonably in pertinent connection with other (consistent) ones.

The *modus vivendi* of the complex being *Homo — i.e. Homo Conscious atque Axiologicus —* is characterized[4] *— more philosophico —* by his poly-axiological activities (*in mente, in anima, in spiritu*) by the Values & Valuations.

Values & Valuations can be classified and considered from many points of view. We consider (and use here consequently) the following " physiological situations " of Values & Valuations in the *Axios Universe*.

I. Autothelical-Values are :

a. Truth   => The work within Truth is *Science* ;
             This hypostasis of Man is named *Homo Sapiens Scientifer,*
b. Good   => The work within Goodness is *Ethics/Morality* ;
             This hypostasis of Man is named *Homo Cogitans,*
c. Beauty   => The work within Beauty is Art ;
             This hypostasis of Man is named *Homo Aestheticus,*
d. Love   => The work within Love is Passion[5] ;
             This hypostasis of Man is named *Homo Amans,*
e. Wisdom   => The work within Wisdom is Philosophy ;
             This hypostasis of Man is named *Homo Philosophans,*
f. Sacra   => The work within Sacra is Religion[6] ;
             This hypostasis of Man is named *Homo Religiosus*[7].

The autothelical values are primordial, *i.e.* are, *ex definitio* :

cardinal values : Truth, Good, Beauty,
integrating values : Love, Wisdom and Sacra.

Culture is the ensemble and the dynamics[8] of Autothelical-Values. The Cultures (created, produced and consumed) by some significant human communities are realized (generated, developed, metamorphosed in *curriculum Historiae*) in the matrices of the culture, e.g. the matrix of culture : the Carpathian-Danubean-Pontean Space.

II. Mean-Values

Mean-Values are very important because the Autothelical-Values are realized through them. The Mean-Values are numerous[9], and numerously are the respective activities ; These hypostases of Man are named :

4. It is an axiological definition of Man and Mankind.
5. The Eros, familiarity, friendship, charity.
6. The major creeds, the beliefs, the Credo-s.
7. With the " levels " Pious and Fides.
8. We cannot insist here on these complex activities of " Man " : the different relevant classifications, the crossing effects, the inflections and metamorphosis, the various examples, etc.
9. Expressed *more humano*, i.e. in human terms. Philosophy and Religion express the cosmic character of Man : *Homo Cosmicus*.

a. Homo Agens
b. Homo Politicus
c. Homo Laborans
d. Homo Economicus
e. Homo Didacticus

f. Homo Informaticus
g. Homo Ecologicus
j. Homo Technicus-Technologicus
k. etc.

Civilization is the ensemble and the dynamics of the Mean-Values. Civilization of a significant community is realized (*idem*) in an alveolus of civilization, e.g. the alveolus of the civilization : the Carpathian-Danubean-Pontean Space.

AN AXIOLOGICAL STATUS OF *HOMO TECHNICUS-TECHNOLOGICUS*

Civilization & Culture express the social character of Man. Therefore, we define Man with all his Values & Valuations using the expression *Homo Humanus* (an accepted pleonastic syntagm enforcing the meaning).

The complexity of the spiritual and material entity named Man — *Homo Axiologicus* (*in potentia, in actu*) — is very wide : there are the interactions between values (" joints ", " sums ", " products ", " interferences ", " inductions ", " synergies ", etc.), and, *ipso facto*, many relations between Civilization and Culture ; e.g. the *modus vivendi et operandi* defined (*in aesentia*) as :

Homo Sapiens Scientifer & Homo Cogitans & Homo Aestheticus & Homo Philosophans & Homo Religiosus & Homo Politicus & Homo Economicus & Homo Technicus & Homo Didacticus & Homo Agens, etc.

Therefore[10], all the relevant activities of Homo Technicus-Technologicus (inevitably, specifically) have also extra-technical tints[11]. But — *ex definitio* — the *modus vivendi et operandi* of Homo Technicus-Technologicus (*stricto-sensu*) and his works (actions, trends) belong to Civilization and not to Culture. The Technical-Technological Value is, therefore, a Mean-Value and not an Autothelical one.

A DEFINITION OF THE FEATURE EMINENS

The eminency of the Homo Technicus-Technologicus is, *ex definitio*, determined by two virtues :

Competence

Gifted engineers, professionally well trained, with creative abilities, relevant technical and social experience, promising training potential, etc. — this kind

---

10. In particular, but very relevant for this scientific-philosophical approach.
11. All the relevant actions of Homo Technicus-Technologicus surpass the strictly technical level.

of good specialists believes with lucidity in the " Romanian resources " (educated workers, skilled craftsmen, experts : organizers scientists, innovators, inventors, etc.) ; they accept many duties showing passion for their profession and committed to their duties, they accomplish a lot of difficult tasks with patience, energy, lucidity, objectively, responsibility, having as their aim (primarily) the social (trans-individual, by group) progress ;

Morality

Tenacious and optimistically, with high professional deontology, patriotic feelings, etc. — this kind of honourable citizens believes that there are enough " forces & energies " [material, spiritual : ethical available funds (of his soul ; of many fellows of the society)] to pass (with inexorable prejudices and pains ; with dramatic sacrifices) this period of history (transient ; with huge individual and social problems) but without any significant corruption of their willingly accepted " moral code " ; Cautious, these honest, committed people feel and detest the mafioso forms of the plats due to the nature of the " real-socialist " regime (actualized by the behaviour of nomenclatura, majorities of aparatciks, and of some employees, expectants, etc.) what exerted their influences (pressure, temptations, perversions, embezzlements, etc.) continuously at all levels of the " totalitarian society ".

There have been some " politicized " engineers, competent persons who were authentic patriots and good keepers of *minima moralia*, living and acting within the frame of the " communist regime " ; These citizens of a " totalitarian state ", governed by specific laws, rules customs of " real-socialism ", were obliged to accept many embarrassing situations (ambiguous, conjectural, professional co-operation with some members of *nomenclatura*, *aparatcik*-s, officers of the Intelligence Services, members of their Secret Political Police - *Securitate*, etc.) in order to save his personal main interests (life, family, profession : information, achievements) and what they consider that could be saved their activity from the society corrupted by many crises (in the field of economy, technique-technology, information, etc.)

A CHARACTERISATION IN AESENTIA OF THE CONSIDERED PERIOD
OF HISTORICAL TEMPUS

The historical period[12] considered in our study is *expressis verbis*, very short (on the historical scale) but very full in social phenomena and processes

---

12. The Temporality is a fundamental form of the Existence (material, spiritual), expressed (*in mente ; in verbum*) by the concept named Time : a) *Cronos* is the Time of the physical, chemical, biological unconscious Existence phenomena and processes of relevant magnitudes (events) ; b) *Tempus* is the Time of psychical and social Existence, actualized in phenomena and processes of relevant magnitudes (events) : this time is " charged " by individual and social feelings, wiliness, trends and other vivid forms of human experiences.

of relevant magnitudes (events) : *i.e.* it is a huge poly-axiological human experience[13].

This extraordinary complex period included two stages :

The stage between *ca.* 1st of December 1989 and *ca.* 15th December 1989, that is before the downfall of the (original) communist dictatorship in Romania (*i.e.* the Ceaucist authoritarian regime). During these days many people (technicians-technologs ; etc.) experienced : despair, anxiety, restlessness, disappointment, impatience, etc.[14] ;

The stage between *ca.* 15th December 1989 and *ca.* 1st of February 1990, that is the stage of the greatest poly-axiological mutations (political, economical, ideological, etc.). During these days, many people experienced : hope, illusions, exaltation, utopian dreams, etc.

SOME RELEVANT FEATURES OF *HOMO TECHNICUS-TECHNOLOGICUS EMINENS* DURING THE CONSIDERED PERIOD

We intend to present in this study only a few interesting features (hypostazis, etc.) of *Homo Technicus-Technologicus Eminens* who, by His/Her activities belong to Romania (society ; civilization & culture)[15]. These selected *casus* express various " point of views " about individual and social (essential, principal, important, noticeable) aspects[16] of the Techno-sphera (and outside of this complex poly-axological world) and — we believe — that the researches of these situations — finalized in remarks, sentences, judgements, studies in articulated intentions, attempts, schemes, projects, plans, activities[17] — will be highly interesting in the future[18]. The whole of these (considered) aspects made a (kind of) *Teatrum Mundi* of Homo Technicus-Technologicus for this period, with many " limit-experiences ".

Hypostasis of Homo Technicus-Technologicus as Homo Laborans Agens

Homo as the class of persons who outline schemes, projects, plans and other activities which define him/her as a relevant ergonomic factor of society, maintaining the reasonable profitable production at that particular " fraction " of History in different fields [mining, metallurgy, chemical indus-

13. We believe that the comparison with a termo-dynamical transformation of phase is pertinent ; this comparison legitimates the methaphora (syntagm) " historical-dynamical transformation of phase ".

14. It is only a brief enumeration of some main feelings of people living in Romania, in this critical complex period of Tempus the Mutatio.

15. *i.e.* the technicians-technologs who are the citizens of this country, or strongly pluri-axiologicaly attached (in mente, in anima) with these men.

16. Modus vivendi-s.

17. Ex ovo, status nascendi.

18. *i.e.* in mutatio, and later in post mutatio : in transitio.

try ; transport means (tractors, wagons, aeroplanes, ships, etc.)] and to enlarge some specific fields (electrotechnics, energy savings, technical cybernetics, automation, etc.) ;

Homo as the class of people who really want to take decisions concerning the activities of some important enterprises (mines, metallurgical plants, chemical combined units, electrical trusts, power electric systems, factories, workshops etc.)[19] to limit (close, metamorphose) enterprises which gained fame for the damage caused to the environment.

All the members of the social- axiological category *Homo Technicus-Technologicus Laborans-Agens Eminens* are, *ex definitio*, *Homo Ecologicus* : *i.e.* *anti-Stultus*, *anti-Entropicus*, *anti-Poluans*, *anti-Aliens*, *anti-Destructor*, *anti-Neccans*.

Many eminent engineers have performed some social activities by which they have tried to present realistically to the decision-making bodies, to members of intelligentsia and to " common people " the " non-ecological " situations, and (if, possible) diminish the proportions of some disasters, or to improve (*idem*) the quality of (natural & humanized) " milieu ".

They have prepared different plans (schemes, effectively projects, etc.) and programs, with an explicit ecological character (with a possible application in the future, *i.e.* in a not extremely long period of historical Tempus) : cleanings (physical, chemical, biological, psychical, social) of many areas of the country, improving some degraded media (*idem*) ; They considered these plans, projects, programs :

achievable (more or less), *in mutatio*, later, with relatively small expenses, with the existent technical means, with the existing working persons (also by adequate participation of the population ; not all these programs (with priority urgencies) were really achievable by " own means " *i.e.* by the technical-economical-ergonomically means of several social-units existing in Romania (the state as a " whole ") in the period of historical Tempus named by us *mutatio* ;

preliminary, which must be included later in comprehensive plans and programs of complex " sanitation " (*idem*) of degraded (natural-humanized) environment, which will be conceived and rapidly put in application *in processus Historiae* in various *modus-es* (with an important participation of different trans-national and large international " social-factors " (some of

19. Like the coal mines, in the area Valea Jiului (the lignite mines of the intra-Carpathean depression of Jiu Valley), the mines of lead and another non-irons ores in the country Baia Mare, the radioactive ore mines (Uranium and seldom elements : Titan, Vanadium, Molibden, Bismut, etc.), Băiţa Bihor, Ciudanoviţa, Crucea Piatra Neamţ, the chemical combined factories (Copşa Mică, Piteşti, Făgăraş, Victoria, Oneşti, Săvinesti, etc.), the oil fields, the atomic plants (in progress : Feldioara, Halânga ; in construction : the nuclear electrical plant Cernavodă), at the Danube Delta, lake Tekirghiol (which contains excellent mud with curative properties), the Black Sea, etc.

" traumas " of environment has trans-national, inter-national/global sources, and consequences).

They have discussed (directly, less directly) with some (real) " decision-ers "[20] some of the (really) decision-makers have known (*grosso modo* ; with some details, etc.) the actually *status-es*[21] only with a shortage of professional knowledge ; The results of these discussions with the " officiants " of the upper part of the social pyramid of Romania had very low efficiency ; In some cases the *eminens* were placed in difficult (individual ; social) positions[22] and their (professional, psychical, moral etc.) capacities for acting pro Technica-Techno-logia were put to severe proofs[23].

Hypostasis of Homo Technicus-Technologicus Eminens as Homo Cogitans

Homo as the class of persons who decide not to leave Romania ; They have patriotic and moral motivations[24] (feelings, creeds, reasons etc.) for staying in Romania[25] because they feel that " something " is going to change in the World and, in Romanian society too[26] ;

Homo as the class of persons who have decided to leave Romania ; They are competent and want to leave their country to improve their career opportunities to acquire better information, to come up to the level of the deci-sion makers and in this much stronger position to become distinguished per-sons in Politics of Science, to penetrate into establishments of powerful, democratic states, to prosper ; They are decided to keep contacts with the

20. Some Eminent were in the staffs of some enterprises but (*ex definitio*) the real power of decision was in that epoch very limited.

21. Phenomena, processes, basic conditions (resources, workers, specialists, etc.), perspectives of developing (engagements, problems, plausible respectively, probable), expected consequences, etc.

22. Criticized, advertised, pushed to the " outskirts " of the society, damned, exiled, eliminated.

23. The tenacious eminent agens persevered in their missio by making some plans considered by them adequate to be actualized in the desired multi-axiological mutatio.

24. They are different from the " precise ", " frozen ", " practical " businessmen, and different from " romantic " persons, who are dominated by utopias (*ex definitio* don't belong to the category of Eminent) ; They know the position of Socialist Republic of Romania with relevant exactitude and in the " concerts of states " express their own opinion in different (quasi-equivalent) manner ; now Romania is no more " the Belgium of Eastern Europe " as it was before the First World War and between the two World Wars ; Bucharest is not even le petit Paris, etc.

25. This type of Homo belongs to (moral) Eminent because he is decided to remain at his East European status that means he will be an Eminent for the Carpathian-Danubian Space Civilization & Culture ; Grosso modo, *in aessentia* (from the official information, from the free information's obtained by different ways, from their own investigations : reliable information's, judgements, evaluations, etc.

26. They know with relevant exactitude the complexity of the lives of many emigrants : of emi-nent, of another men (refugees : political, economical), the lost of rights of citizenship, the struggle for " naturalizations ", the complexity of the meanings of the words visitors, unemployment, guests, tourists, passengers, travellers, asylum, Diaspora, segregation, nostalgia, relations, friends, family, national language, tradition, roots, etc.

country and to develop joint programs[27] (Euro-Romanian or American-Romanian, etc.), partnerships, companies[28].

Hypostasis of Homo Technicus-Technologicus Eminens as Homo Politicus

Homo as a class of persons who give up Politics, because they believe that the policy of the existing " political class " (the rulers, the next ranges of the hierarchy) cannot really solve the problems of Romanian society ravaged by many forms of crises[29] ; Some of them have adopted a passive protest[30] ;

The (small) class of persons continue to criticize the official politicians by their professional competent arguments ; Some of Eminens were punished by the political authority[31] ;

The class of persons which continue to maintain some professional relations with the influent politicians and continue their technical-technological activities (in which some political characteristics are not negligible[32]) ; They believe that a multi-valenced transformation of the society (economy, politics, etc.) is necessary and, thus, inevitable, but they consider (conceive ; desire ; image) this ample transformation as a complex (unequal, non-linear, not " absolutely " complete, not in totality abrupt, etc.) social process in which many existing technical-units (giant enterprises etc.), methods and procedures (socialist non rigid planing, competitions, patents, patterns, etc.) can be modified[33] — (reconstructed, re-adopted ; sponsored by private societies, added by the state ; energized by some applicable measurements for social protections of the employees, workers, technicians, experts and ennobled by drastic and anti-

---

27. See footnote 25. In Europe, in the " soviet-socialist " block, in the relations of super-powers (political, economical, military, technical-technological, etc.), in the long range process of Euro-unification, in the neighbouring states from the (in the Democratic German Republic : the " penetration " of the " wall of shame " in Berlin ; in Cehoslovakia Republic : the " velvet revolution " ; in Hungary, in Yugoslavia : the post-Tito regime ; in Bulgaria : the " revisal " of Jivkov authoritarian regime ; in the Soviet Union : the processes named perestroika and glasnosti ; in Moldavia : the resurrection of national and civilian feeling, etc.)

28. Even at the beginning of December 1989, and *a fortiori* later ; Changing (not in reforming, not in Revolution) ; changing *in mutatio* ; Euro-Romanian unites, American-Romanian unites ; This type of Homo Technicus-Technologicus belong, *ex definitio*, to (moral-pragmatically) Eminent because such persons tries to contribute to the development (progress, extension, etc.) of the Carpathean-Danubean-Pontean Civilization & Culture.

29. A small number of eminent have criticized (directly, in allusions, by their examples) some politicians (*apparatcik-s, politruk-s, culturnik-s*, advisers, organizers, executants, etc.) for some of their activities in which they have evidenced many stupidities, negligence (sabotages, espionage, etc. ; In private discussions, in political meetings (of the organized units of the unique party (the P.C.R.- Romanian Communist Party), in professional meetings, in some lecture, etc.

30. Real apparent resignation, expectations, minimal engagements for the orders received from the leaders, maximal detach, isolation (refuge in arts, in theoretical studies and projects, etc.), preparation for the desired social mutatio.

31. Advertised, replaced, degraded, judged, terrorized, persecuted, sacrificed ; Many engineers really resigned and others committed suicide ; this unhappy men of flabby character can not be included in the category eminent.

32. Military industry, policy of intensive export, political-economical transactions, etc.

33. Not destroyed, partially changed (reduced, divided ; with their programs, products, tools, workers, engineers, reaching a high level of perfection, etc.)

bureaucratic procedures and behaviours) — in such a manner that they will become, in due course, specifically, profitable[34] ;

The class of peoples, which believes in some " socialistic values " adequately, transformed : in different forms of new social issues, which can co-exist together and with new capital-issues ; The spectrogram of technical-polit-ical activities[35] (creeds, ideas, opinions, trends, etc.) is very complex and in-teresting : pro-Technica, pro-Didactica, pro-History ; We cannot expose here this spectrogram[36] ;

The class of peoples, which are Euro-union supporters[37] : which believe that the new Romania will be included in this complex structure edified in two-three generations ;

Homo as a class of persons which are great humanists.

Hypostasis of Homo Technicus-Technologicus Eminens as Homo Didacti-cus-Militans

The class of persons who believe in the efficiency of didactical activities pro Technique and pro Science[38]. they believe in a quick develop-ment of computer aided design and teaching[39] ; They hope and argue that com-puter science will penetrate fast enough to help a few steps in the develop-ment ; they intend to publish updated information[40] and original research (ideas, results, proposals, projects) ; They claim that the development of edu-cational activities is not illusion but a real social force ;

The class of persons who want to participate in the development of effective ecological projects, with existing means or also with more sophisti-cated ones[41] ;

---

34. For local interests, on national scale ; Some of their products can be excellent, and thus resist in the competition of foreign markets.

35. Political activity is necessary ; Politics is not equalled with *politikia-politicianismus* (*i.e.* intrigues, traps, felonies, etc.), the *stricto sensu* non political attitudes in relevant situations (limits, etc.) are a utopia, a dream, a limit.

36. L. Sofonea, E. Helerea, " The status of Homo Technicus-Technologicus Politicus Agens Eminens in the limited situations. The case of Romania in the period *ca.* December 1st 1989 - *ca.* February 1990 ", *Studii de Istoria Stiinţei şi Tehnicii*, Fasc. 3, Braşov, 1998.

37. A new structure (political, economical, culture & civilization *Unio*) : in diversity, the axi-ological common denominator is essential ; This *Unio* is conceived as a " union of homelands " ; The process of integration will be performed using several " gears " and " accelerations " ; The USA is not a model but a relevant example of *Unio*.

38. *Ex definitio*, all kind of Homo Eminens (Technicus, etc.) are Homo Ecologicus ; Homo Technicus.

39. In the social conditions of the " totalitarian regime " many young and old people living in Romania find a spiritual refuge in studium : they have studied, using the specific information means at hand sciences (of Nature, of Society), techniques, literature, etc.

40. The intensive and extensive penetration of computers of different performances) in research, in a lot of technical actions (projects, management, marketing, etc.) and teaching — which is the emblematic sign of Homo Significans-Informaticus — occurs also in Romania but relatively late and slow in comparison with other countries (USA, Canada, Japan, West and Central Europe, etc.)

41. The lack of accurate advanced information was for intelligentsia one of the tributations of the " real socialist regime ".

The class of people who want to develop preservation projects ; they work to identify and to mark certain technical means — *artefacts* — which are currently operational but which will be put rapidly out of order for several reasons : obsolescence, pollution, high levels of risk for users, etc.[42]

Hypostasis of Homo Technicus-Technologicus Eminens Reconstructor

Some eminent technicians-technologists who appreciated History of Technique planned[43] about the following :

Appropriate " social valuation " of some technical artefacts, old and remarkable[44] ones, which still existed[45] in different technical units[46] ;

Appropriate " social valuation " of technical artefacts, old and remarkable, which didn't exist any longer in " technical units "[47] ;

Planning and arrangement of natural & axiological curratorium [48].

---

42. The ecological catastrophes were very frequent in the " real-socialist " regimes : The Copşa Mică chemical plant, the Anina-Crivina bituminous shale excavations, the Valea Jiului smoggy area, the dusts of Hunedoara, Reşiţa, Zlatna, the Uranium mines (Stei-Băiţa, Ciudanoviţa ; Crucea on Bistriţa Valley in Piatra Neamţ country, nuclear chemical plants (Feldioara, district of Braşov ; Halânga, district of Mehedinţi), etc.

43. These plans weren't dreams, they were " organized dreams " (*i.e.* conceptualized, articulated, feasible *in potentia* ; — the authors had used much information (accurate : technical data, plausible facts, etc.) ; some of them had different final terms of execution ; some had even more versions of which their authors hoped/thought that they will turn (immediately after the expected mutatio ; later) into vivid designs (technical designs created by themselves and by other specialists) based on multiple studies and on pluridisciplinary and interdisciplinary serious studies and scenarios) which will (in due course) be developed (mise en oeuvre) and materialized.

44. Having multiple co-relations : vicin-al, region-al, poly-regional, national, trans-national, European, inter-national, planetary.

45. In exhibitions, collections, *vivum museums, studium.*

46. Accomplished through : identifications, studies, retrievable, restorations, preservations, turning into museums ; Some of them were stopped (*i.e.* were not in action) but not disintegrated ; Some were not destroyed, others were sent to be destroyed but their elimination from the technical units was not started/finished ; Some were partially disassembled, modified or forgotten somewhere, etc.

47. They were not melted, transformed into waste ; there still existed some preserved waste matter, some parts were altered, some parts were intact, etc. The contempt for *museisations* of the technical items was a typical attitude of the arrogant communist managers, and rulers ; Some were in operation : they operated with relatively low efficiency, being obsolete (physically and " morally ") but they were not yet destroyed.

48. The " natural & axiologic curratorium " is (a material and spiritual) system, *i.e.* a complex social activity in/through which important axiological assets are preserved and actively used pro-Societatis (Technique History vestiges, etc.) there are important axiological assets in a poor situation : a) have some similar characteristic — or quasi-similar — systems and activities designated by the terms : natural park, natural and axiologic park, natural reserve, museum ; b) have important distinctive features compared to the aforementioned : there are people having a specific " social life " ; Which can be simultaneously done with " the currator-ization " : they are activated by social activities carried out by these complex systems ; Specific care : sanitation, currator-ies, etc. ; L. Sofonea, " Arrangement of a natural & axiologic Carpathean park ", *Proceedings of the first Congress of Social History*, Amsterdam, 1996. The organizers of the second edition has no more given the opportunity to present in this appropriate forum the new considerations concerning this concept, with detailed examples.

Several designs of this type of projects have been made and many multiform " axio-ecological " designs were identified as feasible in reasonable Tempus terms.

First example : *The Danubius - Banatiensis Curratorium.*

It is advocated that in this system[49] (also seen as a complex social activity) the following areas be axiologically and environmentally reconstructed :

Baziaş rail way station and harbour, where the Danube enters in Romania ; Oraviţa-Baziaş rail way was inaugurated in 1850, being the first railway of the Carpathian-Danubean-Pontean Space ;

Oraviţa-Anina Alpine route rail way ;

Uranium mines from Ciudanoviţa, having a complex exhibition [radioactive ore search, discovery and exploitation in Carpathian-Danubean arch (in Tempus periods *ca*. 1930-1946, *ca*. 1946-1957, 1957-1989), at Ciudanoviţa, Băiţa Bihorului, Stei, Sov-Rom Kwartit, etc.] ;

Danube Valley - Porţile de Fier Pass (Clisura-Cazane-Vasskapu-Donauische Eissen-Tore) : the reconstruction of short parts of *via romana*, revival of *tabula Peutingeriana*, of *tabula Traiana* (also a monument) the installation and revival of some artefacts, like the monument and museum dedicated to Orşova history (ancient Dierna), the sinking in 1980 of a great urban — peri-urban area in the artificial lake created by the building of the huge hydro-energetical and navigation system Porţile de Fier, the monument and the (esentialized) museum dedicated to the frontier area of Vârciorova between Austro-Hungarian Empire and Muntenia principality (province Oltenia/Romania), the monument dedicated to the arrival in Romania (in 1866) of Carol of Hohenzolern prince, the monument and the museum commemorating the people who intended to pass the Romanian borders in 1946-1989, in the period of physical and axiological communist holocaust, in order to live a better life (more dignified, safer, more prosperous) in the Free World ; remember of Adakaleh Isle[50], Danube dam and hydro-electric plant (having a small museum), transformation into a museum of the Şimian fortress[51] ;

---

49. Situated in the riveran area of Danube : the left bank (from Baziaş to Drobeta-Turnu Severin, Şimian Ostrov, i.b. Isle) and inside Banat, few kilometres to Anina-Crivina, Herculane Baths, etc.

50. Fortified in the 16-17-18th centuries, a possession of Ottoman Empire ; " integrated " in the system of Turkish Raia-s existing on the territory of Wallachia.

51. Important fortifications were made during the Austrian domination (after the Treaty of Passarowitz) ; the extension of Habsburg monarchy — engaged in the Drang nach Ost Politick — was extended to the Eastern, comprising Oltenia, *i.e. Wallachia minor*, part of principality of Muntenia (Ţara Românescă - Wallachia).

In integrum reconstruction of Vodiţa monastery, close to the original situs, arrangement of a evocative museum of the important religious and cultural contribution of that principality foundation[52] ;

Modernization of Băile Herculane resort[53] : protection and conservation of some technical and curative artefacts in a dedicated museum ;

Protection and conservation of technical artefacts in great villages : Mehadia[54], Topleţ[55] ;

Monuments in Drobeta-Turnu Severin, evoking *tabula Peutingeriana*, the Ioanits Knights diplom, the quarantine station[56] in which must be put eloquently in evidence (in the building from 19th century), completing the Porţile de Fier museum[57], the water tower (monumental water town system), reconstruction of ancient bridge which (2th-3th BC) joined (with pavimentum and another ponts structures) the Daco-Roman shore with the Traco-Moesic — Român shore ; the ruins of the (first in time) roman castrum, the Severus tower, the medieval fortress, basilica, the art museum, the theatre, Halânga heavy water plant, Cerneţ tower (cula)[58], organizing in this Danubean country capital of high level cultural events[59].

Other examples :

Curratorium axed on the two Sarmizegetuza-s [Regia/Bazileia (Kogaionon (?)) ; Ulpia Traiana],
Valea Jiului intra mountain hollow,
Valea Bistriţei,
the Northern Bucovina [hills (*obcine*) and monasteries],
Gorj - Vâlcea - Argeş-Muncel area,

---

52. The Balkan monarchism [the exchanges of priests, erudite, merchants between the Balkan peninsula (the Athos holy mountain, the Kosovo plain, etc.)] and the north Danubean Romanians, the cultural messengers (in the time of Nicodim ; during the first dynasty of Basarab voivods.

53. Which has interesting items : from the period *en vogue* (at the end of the reign of Kaiser Franz-Joseph/Ferenz Joska ciasar, and later).

54. Roman and medieval mines ; relics.

55. Interesting mills : of the type *moara cu făcaie [i.e.* with spoons : an archaic hydro-technical system conceived and used by popular technicians (*magister naturalis*) which anticipate Pelton's turbine].

56. With the topics : Orient - Occident, history of communications, history of medicine.

57. It was inaugurated in 1896.

58. It was designed in 1880 by Elie Radu and Romanian technicians, who won a contest involving also important building and installation companies, contest in front of a committee consisting of state authority representatives ; the constructors were Romanians ; Located in the proximity of this town (*municipium*), capital of the district — judeţul Mehedinţi : the building has an exceptional value (architectural : fortified rustic large houses, with cellar and 2 flats) ; historic : it was a place where the participants (*pandours* ; etc.) in the uprising of 1821 l[ed] by Tudor Vladimirescu operated, the monument is also a part of the natural & ecological curratorium situated in the Mehedinţi - Gorj area.

59. It was a municipium princeps of the province Dacia - Romana (*Felix*) : during the reigns of the Emperors Traian (*Ulpius ; princeps optimus*), Hadrian, etc.

Dobrogea and the Black Sea Coast,

Buzău - Râmnic area,

Prahova - Târgovişte area,

Mărginimea Sibiului,

Făgăraş Country,

Bârsa Country,

Mureş - Covasna - Harghita area (having a very important Hungarian and Sekely minority, being majority in certain areas), etc.[60]

In the eminent's (of these types) conceptions and intentions, these axiological assets (technical, etc.) are endangered, and, *in consecventia*, they require specific protection, and specific " cures ", which have to be done in process Historiae in the organizational frame of the " natural & axiologic curratorium " (complex system and complex social action).

The complex character of such (poli-axiologic) social actions is expressed by :

the remedy of serious (poly-axiological) damages on the " national heritage ",

social valuation of some axiological-assets (components of different degrees of patrimonies),

repairing of some injustices which generated frustration among sensitive consciences (Romania's image, inside for Romanians and for other people ; outside) ; the action put in sensible explicit form the release from " the physical-biotic & axiologic communist holocaust ",

the freedom and the possibility for amplified and diversified social valuations of axiological assets (through reconstruction, museums, etc.)

creation of specific labour opportunities (jobs, stimulation of upgrading, retraining, new training of personnel),

involving *in actio* many people (of various ages, professions, characters, social status) and groups, institutions (quantitative and qualitative), generated solidarity, civic attitudes, patriotic feelings, etc.,

development of tourism (quantitatively ; qualitatively),

development of sports (of some peculiar sports : achieved in these social systems/activities),

development of education (with many aspects : geographical, historical, civic, ethic, patriotic, juridical, sport, etc.),

60. C.C. Giurescu, *Istoria Românilor*, Bucuresti, 1943, 4 vols ; D.C. Giurescu, *Ilustrated History of Romania*, Bucureşti, 1983 ; St. Balan, N.S. Mihăilescu, *Istoria Ştiintei şi Tehnicii în Româ-nia : date cronologice*, Bucureşti, 1985 ; I.M. Ştefan, E. Nicolau , *Scurtă istorie a creaţiei ştiin-ţifice şi tehnice Româneşti*, Bucureşti, 1985 ; N.P. Constantinescu, *Enciclopedia Invenţiilor Teh-nice Româneşti*, Bucureşti, 1934-1946, 3 vols ; I. Simionescu, *Ţara noastră*, Bucureşti, 1938 ; L. Sofonea, " Reconstruirea unor arte facte tehnice de mare valoare istorică din sudul Banatului ", *Lucrările Conferinţei Naţionale a Inginerilor Instalatori*, Timişoara, 1987.

dynamisation (intensification, improving : quantitatively and quali-tatively) of national relationships [" the image of Romania ", multiple interests (the curratoriums — proper places for appropriate and serious discussions, exchanges : of ideas, projects, goods, wills, etc.)],

the power of example (in interior, in exterior),

various by-products (with multiple applications, etc.),

various risks (bankruptcy, stopping of some activities, forms of kitsch (aesthetic, etc.), practical degradation of the norms which are specific (statuary, important) for a natural & axiologic *curratorium*,

ecological aspects, aesthetical aspects, ethical aspects, philosophical and religious aspects.

The very active dreaming eminent technicians and technologists often and intensely thought about what they can do in that historical Time. They were convinced that *post mutatio*, their plans will generate multiple interests and multiple supports will arrived (material and spiritual) which will come true in an appropriate time.

These plans were motivated by the limit situation (psychical/historical) of some Eminent : they were passionate, active, competent, patriots and demo-crats, tortured by the " totalitarian strictness ", frustrated, (often) despaired, indignant (idem), humiliated, enthusiastic connoisseurs/ " fans " of History and Geography (clever analysts of some comparable/quasi comparable situations, investigators of some relevant models of social valuations, etc.), eager " to save what can be saved " and to do something with a social-constructive (multiple) purpose before the expected *mutatio historiae*, and, during its progress.

The " Curratorization " could be done continuously : the inhabitants of this space live (biologically ; spiritually) their lives (non-linear, with multiple changes) and *ipso facto*, being " present " there, they actively participate in this complex social action and they are influenced (specifically) by this process : their material and spiritual existence is ennobled, participating in a *sui generis* school, raising the quality of their lives.

All these considerations certify that for the Homo Technicus-Technologicus Eminens of Romania living in the considered period of Tempus the limited-stage experienced by Him represented a purgatorium and a preparation for the *mutatio* : " ...The time was out of joint... " and " it must be fairly removed in an harmonious place ! "

LE DÉVELOPPEMENT DURABLE AU COSTA RICA OU LA MISE EN PLACE D'*AGENDAS 21* LOCAUX

Patrick MATAGNE

.

Le Costa Rica est une petite démocratie d'Amérique Centrale, caractérisée par sa stabilité politique d'une part, par la richesse de sa diversité biologique d'autre part. Deux raisons qui ont fait considérer ce pays comme une sorte de laboratoire grandeur nature par les associations et les institutions internationales, nationales et locales.

Par ailleurs, une politique écologique ambitieuse a été mise en place à partir des années 1970, puis une nouvelle forme de tourisme écologique s'est développée, avec l'appui du gouvernement.

Depuis la Conférence des Nations Unies sur l'Environnement et le développement (CNUED) qui s'est tenue à Rio de Janeiro en juin 1992, clôturée par un " sommet de la Terre ", des directives en matière de développement durable sont susceptibles de donner une nouvelle impulsion aux modèles déjà mis en place au Costa Rica dans les années 1980. Ces directives, contenues dans l'*Agenda 21*, posent aussi la question des relations entre les pays du Nord développés et les pays du Sud.

LE CONCEPT DE DÉVELOPPEMENT DURABLE ET L'*AGENDA 21*[1]

Le concept de développement durable ou soutenable[2] repose sur la prise de conscience des effets néfastes sur l'environnement, de certains modes de pro-

1. *DIXECO de l'environnement, pour comprendre les échanges entre l'homme et son milieu*, 1995, 73-74. Voir aussi S. Faucheux, J.-F. Noël, *Économie des ressources naturelles et de l'environnement*, Paris, 1995, 239-328 ; J. Tricart, L.-M. Diop-Maes, F. Pesneaud, " Le cercle vicieux du sous-développement peut-il faire place au développement durable ? ", *Annales de Géographie*, n° 579 (1994), 453 et 456 ; I. Sachs, " Le développement reconsidéré : quelques réflexions inspirées par le Sommet de la terre ", *Revue Tiers Monde*, t. XXXV, n° 137 (janvier-mars 1994), 53-55.

2. Vient de l'anglais sustainable development, traduit en français par " développement soutenable ", " développement durable ", parfois par " développement soutenu ", ce qui n'est pas sans introduire un certain flou dans les sens que recouvre l'expression. Dans les documents costariciens consultés, desarrollo sostenible est généralement employé, on trouve parfois desarrollo sustentable.

duction et de consommation[3]. Son objectif est alors de chercher un compromis ou de concilier la nécessité du développement économique avec l'exigence de préserver la nature, mais aussi la vie et les sociétés humaines. Il pose donc la question des limites de la croissance économique et de la recherche de nouveaux modèles de développement.

Sa principale caractéristique est de prendre en compte la variable du temps. C'est pourquoi les modèles proposés s'appuient sur une donnée écologique fondamentale, qui pose que la capacité de compensation, de rénovation ou de récupération des écosystèmes obéisse à des rythmes biologiques et géologiques qui ne doivent pas être transgressés. Ils doivent donc répondre aux besoins des générations actuelles sans compromettre ou affaiblir la capacité des écosystèmes à satisfaire ceux des générations futures. Autrement dit, le développement durable d'aujourd'hui doit permettre celui de demain. Les objectifs économiques ne sont pas seulement quantitatifs, mais aussi qualitatifs. L'économie doit alors introduire la variable écologique comme un facteur de développement, tandis que l'écologie doit intégrer les lois de l'économie.

Au cours de la Conférence de Rio, qui a réuni 107 chefs d'États ou de gouvernement, plus de 3.000 Organisations Non Gouvernementales, 9.000 journalistes, 30.000 participants se sont alors penchés sur une vaste question. Comment arrêter la dégradation de l'environnement tout en permettant le développement des collectivités humaines et notamment celles des pays les plus pauvres ?

Cinq propositions ont fait l'objet de discussions et de débats, parfois d'affrontements entre les pays du Nord et ceux du Sud :

- une " charte de la terre " devenue " déclaration de Rio ", qui jette les bases d'une " gestion écologique rationnelle " de la planète. Le texte est généralement jugé consensuel, mais difficile à traduire en actes.

- une convention sur la forêt. Mais son extrême prudence est la marque de la pression des producteurs et des consommateurs de bois tropicaux.

- une convention sur la biodiversité, que les pays du Nord et du Sud ont voulu peu contraignante, car les premiers souhaitent protéger leurs sites industriels et les seconds entendent conserver leur souveraineté nationale sur leur biodiversité.

- une convention sur le changement climatique, qui se réduit à un simple engagement moral sans obligation précise de limiter l'émission de gaz à effet de serre.

- un programme d'action en faveur du développement, appelé *Agenda 21*. Chaque pays s'engage alors à intervenir dans un domaine (pollution de l'eau, exploitation forestière, etc.) et à en assurer le financement. En fait, la plupart

---

3. Bruntland, *Notre avenir pour tous*, Montréal, 1988.

des signataires se sont contentés d'augmenter leur contribution à divers fonds internationaux, dans des proportions inférieures à celles qui étaient prévues.

Pourtant, l'*Agenda 21* exige une meilleure prise en compte de l'environnement dans les politiques, dans la planification, et un renforcement du cadre juridique. Cette première règle est reprise de la déclaration des droits de l'homme à l'environnement de la conférence de Stockholm, le 16 juin 1972, avec des termes différents. Elle implique l'existence de biens communs à l'humanité et d'un patrimoine mondial naturel et culturel. Cette mondialisation doit être conciliée avec le maintien du droit souverain des États, sous réserve de ne pas causer de dommages à l'environnement des autres États et aux espaces communs (les océans en particulier).

La deuxième règle, retrouvée dans l'article 130R du Traité de Maastricht, pose la nécessité d'intégrer l'environnement dans toutes les politiques sectorielles. Il s'agit d'aider au développement tout en protégeant l'environnement. Cette aide passe par la réduction de la pauvreté et des modes de production et de consommation non viables, qui sont de la responsabilité des pays développés.

La troisième règle est celle du " pollueur-payeur ".

La quatrième règle est dite de non-discrimination. C'est une règle de bonne conduite écologique entre États. Elle vise les problèmes transfrontaliers et de commerce international (transport de produits dangereux). Autrement dit, il ne faudrait pas imposer à un État voisin des règles de protection de l'environnement moins favorables que celles qu'on applique aux nationaux. Mais les exportations de déchets ou de pesticides interdits à la consommation dans les pays producteurs restent possibles dans certaines conditions. Il est clair que l'environnement devient un enjeu dans la guerre commerciale.

La cinquième règle pose que l'environnement est indissociable des principes de démocratie participative et de solidarité. Elle impose un devoir d'information, de consultation, d'assistance en cas de catastrophe.

La sixième règle impose l'adoption d'un " principe de précaution " ; selon lequel il serait préférable de ne pas agir ou de renoncer à une action non-maîtrisée, notamment dans des domaines controversés comme le nucléaire, la disparition de la couche d'ozone ou les manipulations génétiques.

Enfin la septième règle concerne l'anticipation, la prévention des causes de la diminution de la biodiversité, grâce à des études d'impact (d'où la promotion de programmes d'inventaires), à la mise en place de normes écologiques, à la gestion écologiquement rationnelle, en particulier celle des forêts, dans lesquelles l'" érosion biologique " est importante.

La conférence de Rio, critiquable à bien des égards, a toutefois posé les bases d'une réflexion théorique et a permis de maintenir le dialogue entre les pays du Nord et ceux du Sud, malgré les tensions qui se sont exprimées au cours des débats. L'ensemble de ces recommandations implique de réfléchir à la nature des relations Nord-Sud, aux orientations politiques des États, aux programmes de développement régionaux et locaux, notamment dans le domaine de la gestion des forêts qui sont les principaux réservoirs de diversité biologique, dans celui de la gestion des villes et de leurs nuisances. Elles posent également le problème de la formation des " décideurs " en matière de développement[4].

Au Costa Rica, la question de la protection de la forêt se pose de façon dramatique car la déforestation y est particulièrement sévère depuis les années 1940 (document 1), en raison de la mécanisation de l'agriculture et de l'extension de l'élevage. Pour répondre aux besoins du marché international, surtout à la demande de viande bovine dans les années 1970, l'État costaricien a alors réservé jusqu'à 50% des crédits agricoles à l'élevage. Actuellement, les dix grandes formations végétales forestières originelles se réduisent à quelques lambeaux, soit 25% de la superficie en 1987[5], dont près de 19% sont protégées[6] (document 2).

C'est ainsi que le Costa Rica abrite encore plus d'un demi-million d'espèces animales et végétales[7], soit 5% à 6% de l'ensemble de la planète, sur un petit territoire qui ne représente que 0,03% des terres émergées ! Cette biodiversité[8]

---

4. En France, L'université d'Orléans vient de créer un DEA (1995), Environnement : Temps, Espaces, Sociétés (ETES), sous la responsabilité de Jean-Paul Deléage. Son projet scientifique est de former une communauté capable de traiter les problèmes d'environnement soulevés par le développement des sociétés contemporaines. L'environnement étant une notion transversale, le projet pédagogique du DEA est interdisciplinaire. Il vise à approfondir les connaissances scientifiques et à repérer les concepts essentiels des études sur l'environnement, à intégrer ses différentes échelles (temps, espace, niveau global, régional, local), à apprendre à travailler en équipe en vue de recherches interdisciplinaires. Un DEA Economie de l'Environnement et des Ressources Naturelles existe aussi à l'Université de Paris 1, Panthéon Sorbonne. Par ailleurs, les nouveaux diplômés dans le domaine de l'aménagement du territoire sont sensibilisés aux questions environnementales.

5. 8% sont administrés par le Service des Parcs Nationaux, 8% par la Direction Forestière, 2,5% par la Direction de Vie Sylvestre, 5% par la Commission Nationale Indigène, 1% par des organisations privées comme le Centre Scientifique Tropical et l'Organisation des Études Tropicales. Le reste se trouve sur des propriétés privées et des aires récréatives administrées par le Ministère de la Culture, de la Jeunesse et des Sports ou par les municipalités. Sources : A. Bonilla Duran, T.A. Meza Ocampo, *Problemas de desarrollo sustentable en América Central : el caso de Costa Rica*, 1994, 90.

6. La loi forestière du 28 juin 1990, titre 2, chapitre 1, article 35, les définit comme suit : réserve forestière (exploitée pour le bois), zone protégée (maintien du sol et protection des bassins hydrographiques), parc national (protection des beautés naturelles, de la faune et de la flore, accessible aux chercheurs et au public), refuge national de vie sylvestre (protection, conservation, développement de la faune et de la flore sylvestres), réserve biologique (conservation et étude de la vie et des écosystèmes sylvestres). T. Meza, A. Bonilla, *Areas naturales protegidas de Costa Rica*, Cartago, 1993.

7. Près de 1.400.000 espèces sont répertoriées au niveau mondial, mais on estime leur nombre à près de 10 millions. Au Costa Rica sont répertoriées : 1.300 espèces d'orchidées, 378 espèces d'amphibiens et reptiles, 846 espèces d'oiseaux, etc. Voir aussi L.D. Gomez, " La biodiversidad de Costa Rica ", dans R. Salazar, J.A. Cabrera Medaglia, A. Lopez Mora (éds), *Biodiversidad politicas y legislation a la luz del desarrollo sostenible*, 27-29 ; V. Solis, " Comentario a la exposicion de L.D. Gomez ", *loc. cit.*, 31.

8. La biodiversité peut être définie comme la diversité des espèces et des écosystèmes, incluant le matériel génétique. J.A. Cabrera Medaglia, " Presentacion ", *Biodiversidad politicas y legislation a la luz del desarrollo sostenible*, loc. cit., 7.

est le résultat d'un concours d'influences d'ordre biogéographique, climatique, topographique et géologique exceptionnel[9].

Pour lutter contre la déforestation et restaurer certaines surfaces encore boisées il y a une trentaine d'années, des communautés tentent de promouvoir de nouveaux modèles de développement.

L'ASSOCIATION GUANACASTÈQUE DE DÉVELOPPEMENT FORESTIER (AGUADEFOR)[10] ET LE DÉVELOPPEMENT DE L'AGRO-SYLVO-CULTURE

La péninsule de Nicoya (document 3) est la région la plus anciennement et la plus fortement déforestée. A la fin des années 1970, la région de Guanacaste est restée sans forêts, dans des conditions socio-économiques catastrophiques[11] à cause de la chute des cours mondiaux du café et de la viande, face à une crise écologique se manifestant par la sécheresse et l'érosion des sols. Sans parler de la pollution due à l'utilisation massive de pesticides[12].

AGUADEFOR est une association fondée en 1986 par un groupe d'agriculteurs costariciens venus du Plateau Central. Le noyau de départ va être constitué de 119 petits et moyens producteurs, des *finqueros* répartis sur 6 cantons de la péninsule de Nicoya. Ils se donnent pour mission d'obtenir que les activités productives soient en accord avec les principes écologiques qui permettent un développement durable. Ils veulent alors développer à la fois des activités de conservation et de production.

Actuellement, AGUADEFOR est le représentant d'une vingtaine d'associations cantonales et de coopératives regroupant environ 5.000 membres. Elle est soutenue par le gouvernement costaricien et par des organisations internationales et locales qui travaillent dans le même sens[13]. Son but est de faire reforester les bassins fluviaux afin de lutter contre la désertification, la sécheresse et l'érosion des sols, de créer des agro-écosystèmes, notamment dans le cas de la culture du café.

Les problèmes qui se posent sont doubles :

- il s'agit d'opérer un changement de mentalité, de faire prendre conscience de la nécessité d'une gestion à long terme d'une *finca* (exploitation agricole).

9. J.-F. Beauvais, " L'origine de la biodiversité au Costa Rica ", *Bulletin de la Société Botanique du Centre-Ouest*, 1997.

10. Entretien avec Juan Rafael Marin Quiros, directeur exécutif (le 7 août 1995). *Perfil* 1994, document d'information d'AGUADEFOR. "Informe departamento desarrollo industrial a la junta directiva de AGUADEFOR " (juillet 1995), rapport de Eduardo Barrantes Quiros. *ACA news*, vol. 2, n° 1 (mars 1993).

11. Alors que le PIB a augmenté entre 1960 et 1980, il a chuté entre 1980 et 1984. Voir A. Bonilla Duran, *Crisis ecologica de América Central*, 1988, 5-6. Voir aussi, Luis A. Fournier, *Desarrollo y perspectiva del movimiento conservacionista costarricense*, 1991, 56-70.

12. Lindane, Paration, Clordano, Toxafeno, Aldrin, Dieldrin, etc., interdits ou restreints dans les pays du Nord.

13. Associacion de desarrollo Forestal Brunca, Atlantico, Pacifico Central, Ligue de Monteverde.

- Il s'agit aussi de résoudre les conflits entre les banques et les petits exploitants.

AGUADEFOR peut alors obtenir des crédits en garantissant leur remboursement et verser au *finquero* le bénéfice d'une plantation d'arbres, avant même de la commencer. Il ne s'agit donc pas d'une subvention mais d'une avance sur recette. AGUADEFOR fournit également une assistance technique[14], une aide à la conception et à la réalisation de projets, une formation, une information et intervient dans les écoles.

Actuellement, 15.000 ha sont replantés et des *beneficios* (des entreprises de café) associent des arbres à leur culture. Ils produisent des revenus annexes (bois, fruits), apportent de l'ombre, maintiennent le sol, servent de brise-vent, apportent de l'azote (grâce à l'emploi de légumineuses comme Erythrina).

A terme, les producteurs doivent avoir la possibilité de s'adapter aux variations des prix internationaux en diversifiant les productions et en proposant des produits de qualité, répondant aux nouvelles normes internationales. Ils développent aussi l'agriculture biologique. Une coopérative agricole (COOPEPILANGOSTA) est un modèle de réussite. Elle produit et conditionne un café écologique, le café *forestal*, qui tente une percée sur le marché européen, grâce au soutien du WWF Suisse[15].

Enfin, AGUADEFOR développe depuis 1994 un programme industriel qui doit permettre une utilisation et une commercialisation du bois[16]. L'association travaille ainsi à acquérir son autonomie vis-à-vis des organismes internationaux.

MONTEVERDE, UN PROGRAMME POUR UN FUTUR SOUTENABLE (Document 4)

L'histoire des communautés de Monteverde remonte aux années 1950, avec l'installation de quelques familles de Quakers (ou " Société des Amis ")[17]. Quand la communauté arrive, le Costa Rica sort d'une guerre civile (1948) et le nouveau président José Figueres vient d'abolir l'armée[18]. Les Quakers sont alors séduits par le petit État démocratique et pacifiste. Ils nomment Monteverde la région qu'ils choisissent pour son climat agréable et sa forêt tropicale.

---

14. L'association se fait rétribuer en prélevant 10 à 12 % des bénéfices obtenus grâce à son intervention.

15. Entretien avec Gerardo Quesada, producteur de café, et Luis Salazar, comptable de Coocafé, en août 1995. Voir aussi O. Van Bogaert, " Foresta, un concept social et écologique : la nature dans votre tasse de café ! " et " Coopepilangosta lave plus propre ! ", *Panda Nouvelles* (1993), 3-6.

16. E. Barrantes Quiros, " Informe departamento desarrollo industrial a la junta directiva de AGUADEFOR " (juillet 1995). Pour une étude plus générale, J.K. Boyce, A.F. Gonzalez, E. Fürst, O. Segura Bonilla, *Café y desarrollo sostenible, del cultivo agroquimico a la produccion organica en Costa Rica*, Heredia, 1994.

17. M. Mendenhall, " Monteverde ", *Canadian Quaker Pamphlet Series*, n° 42 (Argenta, 1995).

18. C. Meléndez, *Historia de Costa Rica*, San José (Costa Rica), 1991 ; O. Dabène, *L'Amérique latine au XXᵉ siècle*, Paris, 1994.

Ils y créent un centre, point de départ d'un projet de développement qui prend forme dans les années 1980 avec un réseau d'institutions. En particulier avec la Ligue de Conservation de Monteverde fondée en 1986, qui impulse et oriente diverses activités dans les domaines culturels et éducatifs, centrés sur la famille, la musique, le théâtre, l'architecture soutenable, la production écologique, etc. La Ligue est soutenue financièrement et animée par des universités américaines, par l'université du Costa Rica et plusieurs organisations[19].

Le projet dont il va être question ici est celui qui a été initié en 1993. Il est nommé bosques en fincas. Il est fondé sur la création d'un réseau de corridors de vie sauvage, que les anglophones locaux appellent green network et les hispanophones enlace verde[20]. Ces corridors permettent de connecter les aires protégées. En effet, des études ont montré que ces sanctuaires de vie sauvage sont, à terme, non viables pour un certain nombre d'espèces animales qui ont besoin de circuler et de migrer. Il s'agit notamment des grands mammifères comme les pumas et de certains oiseaux.

La viabilité des aires passe donc par la création de corridors reforestés dont la largeur et la composition spécifique sont prévues pour rétablir la circulation des animaux, la dissémination des graines, la pollinisation par les insectes, les oiseaux ou les chauves souris.

Le projet pilote (document 5) part de la réserve de Monteverde (A), reliée par un couloir à la Forêt Éternelle des Enfants (2). Ce couloir préexistait et le jeu des achats successifs de terres vers l'Est a permis de le maintenir et de le renforcer. La zone tampon permet de limiter les agressions des écosystèmes de la réserve, leur altération en lisière par la chasse, les pâtures, l'exploitation du bois. A partir de cette unité de base, de cette " matrice ", des corridors interrégionaux peuvent faire le lien avec d'autres unités de base identiques. C'est le cas avec l'aire d'Arenal. A terme, le couloir isthmique pourrait être restauré, rétablissant ainsi la circulation des populations animales dans la direction nord-sud et entre les versants Atlantique et Pacifique.

Ce dispositif peut aussi contribuer à augmenter l'attrait pour les parcs et promouvoir l'écotourisme, soutenu par le gouvernement[21].

Des conseils sont donnés aux propriétaires pour savoir où planter et comment prendre soin des arbres, comment tenir le bétail éloigné des pépinières.

19. Universités de Californie (éducation à l'étranger), de l'État de New-York à Buffalo (école d'architecture), du Mariland (architecture paysagiste), du Kansas (architecture), du Colorado (architecture), de Milwaukee (architecture). University Programs Council for International Educational Exchange, International Nutrition Program for Developing Countries, International Union for the Conservation of Nature, etc. Latin American Center for Feminist Research de San José, Allianza de Mujeres Costarricenses, etc.

20. C. Guindon, " Forest remnants and corridors ", *Tapir Tracks*, vol. 9, n° 1 (mars 1994), 1-2. Enlace verde, *Monteverde Institute*, rapport annuel de 1994.

21. " Costa Rica. Enjeu politique et cas d'école ", *Courrier international*, n° 195 (28 juillet-17 août 1992), 12 ; K. Lindberg, D.E. Hawkins, *Ecotourism : a guide for planners & managers*, Vermont (North Bennington), 1993.

S'ils savent que les arbres contribuent à prévenir l'érosion, s'ils sont convaincus qu'en préservant la forêt ils procureront un abri naturel pour le bétail qui produira alors plus de lait, ils pourront voir les corridors comme une source de profit supplémentaire, comme une aire qui augmente le potentiel de leur terre. Mais beaucoup de fincas sont déjà morcelées, les paysans doivent alors surexploiter de minuscules lopins de terre, pour survivre avec leur famille, dans l'incertitude de l'avenir. A l'échelle du pays, les deux-tiers des costariciens qui sont sous le seuil de pauvreté sont des ruraux[22]. Dans ces conditions, il paraît difficile de faire valoir l'intérêt à moyen ou à long terme d'une agro-écologie. Il est clair que la pauvreté est un frein au développement durable. De plus, au Costa Rica les limites des propriétés sont souvent floues, une même parcelle pouvant même avoir plusieurs propriétaires. Les décisions d'achat pour protection, par les associations, sont alors problématiques.

Conclusion

Le cas Arenal-Monteverde pose un certain nombre de problèmes de fond, qui touchent le concept même de développement durable. S'agit-il d'une nouvelle forme de colonialisme écologique des pays du Nord développé ? Jusqu'à quel point peut-on instaurer un droit d'ingérence écologique des pays du Nord vis-à-vis de ceux du Sud, et dans ce cas, ce droit ne devrait-il pas être symétrique ?

AGUADEFOR, après un débat interne, semble vouloir travailler à son indépendance. L'association est issue d'une initiative costaricienne et cette aspiration paraît légitime. A Monteverde au contraire, les programmes sont impulsés de l'extérieur. Certains, parmi les plus actifs de la Ligue de Conservation de Monteverde[23], ne cachent pas les difficultés rencontrées auprès de la population locale. la " greffe " aurait du mal à prendre !

En fait, ces programmes intègrent l'idée que la biodiversité doit faire partie du patrimoine de l'humanité. C'est le cas de la Forêt Éternelle des enfants, qui appartient symboliquement à des enfants de plusieurs nationalités. Des intellectuels costariciens remettent en cause cette conception, considérant même qu'il s'agit d'une véritable escroquerie[24].

Par ailleurs, certaines orientations prises par le gouvernement costaricien manquent de lisibilité. Derrière les discours conservationnistes officiels destinés à promouvoir l'écotourisme, se cache parfois une toute autre réalité. Des entreprises touristiques s'approprient le mot " écotourisme " sans avoir aucune vocation à le promouvoir et les principaux bénéfices sont retirés par des étran-

22. En 1984, les statistiques officielles ont recensé 16.724 fincas de moins d'un hectare, tandis que 1.722 ont plus de 500 ha. *Programa de desarrollo rural*, Gobierno de Costa Rica, 1994, 19 et 34.

23. Carlos Guindon en particulier, qui travaille à la mise en place du réseau vert.

24. R. Salazar, *et al.*, *Diversidad biológica y desarrollo sostenible*, San José, 1993.

gers. Le gouvernement précédent n'avait-il pas envisagé la privatisation des parcs nationaux et leur cession à des consortiums japonais ?

Le chantage et la corruption pèsent également sur un certain nombre de projets. Il n'est pas rare que des fonctionnaires donnent des autorisations de faire des coupes alors que seule la Direction Forestière est habilitée à le faire. Au sein même de cette administration, des coupes sont officiellement autorisées dans des aires protégées. Pire, certains plans gouvernementaux sont pervertis au point que des forêts naturelles ont été coupées pour y établir des projets de reforestation, qui s'accompagnent d'une aide financière.

A l'heure actuelle, les modèles de développement durable appliqués au Costa Rica sont des solutions locales, dont la durabilité est souvent mise en péril par l'" analphabétisme écologique " qui sévit dans les milieux politiques[25], par la corruption, la pauvreté, les intérêts à court terme, le manque d'indépendance culturelle. Pourtant, il semble évident que le développement du Costa Rica, sa survie matérielle et culturelle, passe par l'application d'une véritable politique écologique.

---

25. A. Bonilla Duran, *Crisis ecologica de América Central*, 1988, 103.

CONTRIBUTORS

Michael C. DUFFY
S.E.A.T.
University of Sunderland
Sunderland (Great Britain)

José M. CANO PAVÓN
University of Malaga
Malaga (Spain)

Elena HELEREA
Universitatea Transilvania
Brasov (Romania)

Eberhard KNOBLOCH
Technische Universität
Berlin (Germany)

Cesare S. MAFFIOLI
Université Louis Pasteur
Stasbourg (France)

Patrick MATAGNE
Tours (France)

Bruno MEYER
E.T.H.
Zürich (Switzerland)

James Gabriel O'HARA
University of Hannover and
Niedersächsische Landesbibliothek
Hannover (Germany)

Hanuš SALZ
Aero Historia
Plzen (Czech Republic)

Konstantin SÉVÉRINOV
Université d'Électrotechnique
d'Etat de St. Petersbourg
St. Petersbourg (Russia)

Véra SÉVÉRINOVA
Université d'Électrotechnique
d'Etat de St. Petersbourg
St. Petersbourg (Russia)

Liviu SOFONEA
Universitatea Transilvania
Brasov (Romania)

Shigeo SUGIYAMA
Hokkaido University
Sapporo (Japan)

Margarita M. VORONINA
St. Petersbourg State University
St. Petersbourg (Russia)

Karel ZEITHAMMER
Czech Technical University
Prague (Czech Republic)

Baichun ZHANG
Chinese Academy of Sciences
Beijing (China)